INTERPRETATION OF ORGANIC SPECTRA

INTERPRETATION OF ORGANIC SPECTRA

Edited by

D. W. MATHIESON

The School of Pharmacy

London University

England

1965

Published in association with
the Royal Institute of Chemistry
by Academic Press, London and New York

ACADEMIC PRESS INC. (LONDON) LTD
Berkeley Square House
Berkeley Square
London, W.1

U.S. Edition published by
ACADEMIC PRESS INC.
111 Fifth Avenue
New York, New York 10003

Second printing 1966

Library of Congress Catalog Card Number: 65-24816

Printed in Great Britain by
Spottiswoode, Ballantyne and Company Limited
London and Colchester

Preface

The discussions of the various spectra which appear in this book formed the basis of seminar sessions at a summer school in organic spectroscopy; this was held under the aegis of the Royal Institute of Chemistry. In these discussions any necessary background of theory has been assumed. For pedagogic reasons, moreover, little or no ancillary information about each unknown example has been given. This was done in the hope that the strength as well as the limitations of each type of technique might become plain. It was recognized that this was an artificial situation; spectroscopists would normally be in receipt of "background" information which would be of material assistance in interpreting a spectrum.

The sections on infrared spectroscopy and nuclear magnetic resonance spectroscopy have been contributed by some of the tutors at the summer school mentioned above; the section on mass spectrometry was contributed by Mr. Alan Quayle and Dr. R. I. Reed.

D. W. Mathieson

London
May 1965

List of Contributors

Nuclear Magnetic Resonance Spectroscopy

J. A. Elvidge, *Department of Chemistry, Imperial College, London, England*

Infrared Spectroscopy

J. K. Brown, *Department of Chemistry, University of Birmingham, England*

K. J. Morgan, *Department of Chemistry, University of Lancaster, St. Leonards House, St. Leonards Gate, Lancaster, England*

C. J. Timmons, *Department of Chemistry, The University, Nottingham, England*

D. Whiffen, *Basic Physics Division, National Physical Laboratory, Teddington, Middlesex, England*

Mass Spectrometry

A. Quayle, *Millstead Laboratory of Chemical Enzymology, Sittingbourne, Kent, England*

R. I. Reed, *Department of Chemistry, The University, Glasgow, Scotland*

Contents

Nuclear Magnetic Resonance Spectroscopy

Introduction

It is assumed that the reader has acquired some knowledge of the principles, practice and terminology of nuclear magnetic resonance spectroscopy.[1] The intention here is to demonstrate how this knowledge may be used to translate the spectrum into terms of an organic structure.

Reference data in the form of tables of τ values and coupling constants (J) are appended (pp. 51–58). The literature[1, 2, 3] contains further information. It seems pertinent to reiterate that interpretation of a spectrum embraces the following: (a) the positions of lines* (the chemical shift—τ value—gives a clue as to the structural environment of the protons concerned); (b) their intensities† (the intensity gives the number of protons responsible); (c) their multiplicity (this indicates the number of protons on adjacent carbon atoms and thereby structural and often stereochemical information).

One must recognize which signals originate from differently bound protons. Thus a single line may arise either from a single proton or from a group of equivalent protons; a group of lines may arise from as many non-equivalent protons as there are lines, or it may arise from *one* component only of a spin-coupled system. In the latter case there will be one or more other groups of lines corresponding to the other coupled components.

When the chemical shift δ in cycles per second (c/s), between two groups of protons comprising a spin-coupled system is large compared with the coupling constant J (c/s), then the spectrum is "first-order" and the analysis of the multiplets follows the simple splitting rule, viz:—

The number of lines arising from a group of equivalent protons $N = n+1$, where n is the number of *neighbouring* protons (equivalent among themselves) which are coupled to the first group.

Thus, for the ethoxyl grouping, $-O—CH_2—CH_3$ (an A_2X_3 system)‡, the three equivalent methyl protons give rise, not to a single signal of

---

* Line positions are calculated from the expression:

$$\tau = 10-\delta \text{ ppm} = 10-\frac{(\text{line position from tetramethylsilane in c/s})\times 10^6}{(\text{operating frequency in c/s})}$$

† Strictly integrated intensities, i.e. areas under peaks. "Intensity" will be used in this sense, within this Section. Line widths vary widely so that the heights of lines cannot be used as a measure of relative intensities.

‡ The conventions of this symbolism are given in Ref. 1, p. 72.

integrated intensity 3, but to a *triplet* (of the same total intensity 3) due to coupling with the *two* methylene protons: in the above "splitting rule", $n = 2$. At the same time the two methylene protons appear as a quartet (because $n = 3$), of total intensity 2. The intensities of the *individual lines* of a multiplet correspond (ideally) to the numerical coefficients in a binomial expansion, $(1+r)^n$. Thus the lines of the methyl triplet have *relative* intensities 1 : 2 : 1, and those of the methylene quartet have relative intensities 1 : 3 : 3 : 1, as shown.

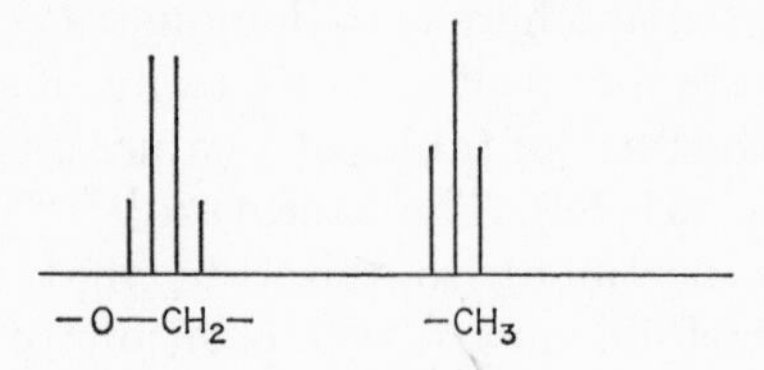

When the chemical shift δ (c/s) between two protons, or groups of protons, of a spin-coupled system, is comparable with or less than the coupling constant, J (c/s), then patterns of lines arise which are not explicable by simple rules and recourse must be had to quantum mechanics for interpretation.

Between these two situations, however, the spectrum is frequently recognizable on a first-order basis although the relative intensities of the lines of the various multiplets are distorted. Thus the pattern of lines produced by the five protons of an ethoxyl group (shown in ideal fashion above) has, in practice, the following appearance.

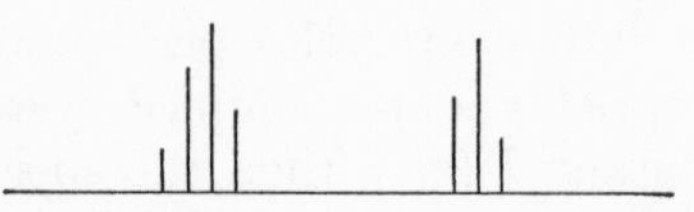

The inside lines of each multiplet are increased in intensity at the expense of the outer lines. This distortion is useful in that its direction within one multiplet indicates on which side the other component lies, i.e. whether to lower or higher field.

It should also be remembered that coupling constants have signs + and − as indicated in Table 6, p. 57.

Proton Magnetic Resonance Spectra (at 60 Mc/s) and their Interpretation

1. Compound $C_{10}H_{10}O$

Solvent: carbon tetrachloride
Sweep width: 500 c/s
Sweep offset: zero

The signal from the added internal reference, tetramethylsilane, is at τ 10 (by convention). The solvent, carbon tetrachloride, gives no signal, of course, so all the other lines arise from the dissolved compound. Calibration of the integral trace* helps in assessing the significance of the various signals. The full rise in the trace is divided by the total number of protons, ten. The result gives the size of a step equivalent to one proton.

It follows that the high-field line at τ 7·75 arises from *three* protons. These are equivalent and are not coupled to other protons, because they give a single line, and so they are probably the protons of an isolated methyl group. Table 1, p. 51 indicates that this might be attached either to an aromatic ring (τ 7·66 for toluene) or to a carbonyl group (τ 7·90 for CH_3CO).

The integral trace further shows that the lines at 386·5 and 403 c/s from tetramethylsilane (TMS) *together* constitute the signal from a *single* proton. The weighted-mean position, the origin position of this proton signal, is at τ 3·4, so the proton is evidently olefinic (Table 4, p. 54). Because the signal is split as a *doublet*, there is coupling to *one* other proton (in the simple splitting rule, $N = 2$ so $n = 1$), and because the lower-field line (i.e. left-hand line) of the doublet is the more intense, the signal from the other proton must be at lower field. To lower-field is a complex pattern of lines arising, as the integral trace shows, from *six* protons. However, two lines within this pattern at 436 and 452·5 c/s are discernable as practically the mirror image of the doublet at τ 3·4. There is then a partially obscured AB quartet in the spectrum:

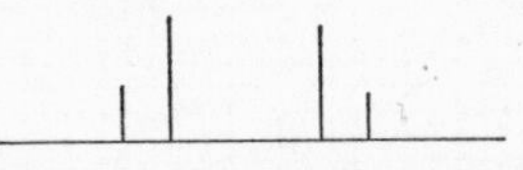

* The steps in this trace give the integrated intensities of the lines. The electronic integrator is operated on a separate occasion from the recording of the absorption signals and so it does not matter if any of the absorption lines have run off the chart paper and appear with their peaks cut off. The integrator does not scan the trace of the absorption signals scribed on the chart paper.

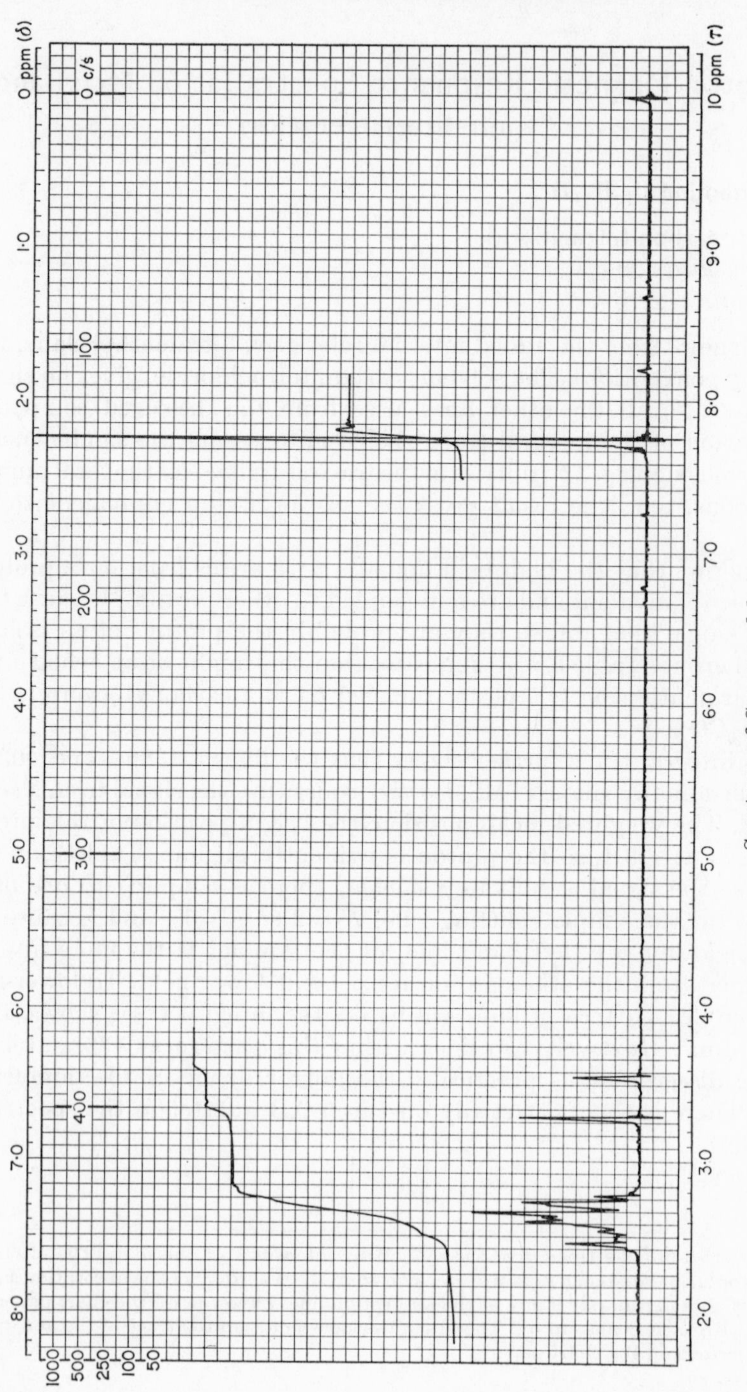

Spectrum of Compound 1.

The lower-field doublet component has an origin position at τ 2·6, a region of the spectrum (Table 4, p. 54) in which highly deshielded olefinic protons and aromatic protons are found. The magnitude of the splitting, $J = 16{\cdot}5$ c/s, indicates that the two coupled protons are *trans* about a double bond (Table 6, p. 57). This *trans*-olefinic link must be isolated in the molecule from other proton-containing groups because there is no further coupling. (This would increase the multiplicity of the proton signals.)

The five remaining protons in the molecule, responsible for the rest of the lowest-field signal, which is centred at 440·5 c/s (τ 2·65), could be assigned to a phenyl group. Table 4 (p. 54) shows that benzene gives a resonance line at τ 2·73 and that an electron-withdrawing substituent would lower this value a little. At the same time, the substituent would probably render the protons on the benzene ring non-equivalent. So the appearance of the 5-proton signal can be accounted for.

These deductions suggest that the molecule is comprised of the fragments

$$C_6H_5–,\quad \begin{matrix} \diagdown \quad\quad\;\; H \\ C{=}C \\ H \quad\quad\;\; \diagdown \end{matrix},\quad –COCH_3,$$

and so the compound is benzylidene-acetone. The phenyl and carbonyl groups at each end of the double bond account for the low-field positions of the olefinic protons, that at lowest-field being the one adjacent to the benzene ring.

The benzylidene–acetone spectrum happens to have a very low noise-level, so that certain features commonly present in spectra, but which are usually obscured by the base-line noise, show up clearly. In the present case, these features comprise the six very small signals which are arranged symmetrically about the sharp methyl-singlet at τ 7·75.

The two outermost, very small absorption signals, at 72 and 199 c/s from T.M.S. are so-called "carbon-13 satellites". Because of the $\sim 1\%$ natural abundance of ^{13}C, about 1% of the CH_3 group in benzylidene-acetone will be $^{13}CH_3$. Unlike ^{12}C, which is non-magnetic ($I = 0$), the nucleus of ^{13}C has a spin of one-half ($I = \frac{1}{2}$), like the proton. The simple "splitting rule" will therefore apply within the $^{13}CH_3$ group, and we deduce that the carbon nucleus causes splitting of the signal from the attached three equivalent protons into *two* lines. The ^{13}C– to –H coupling constant is of the order of 120–200 c/s. In the present case, the lines of the doublet are about 127 c/s apart. The combined intensity of the two lines is about 1% of that of the main singlet (at τ 7·75) which arises from the $^{12}CH_3$ protons.

Sometimes, the satellite signals are more prominent, as in the spectrum of dioxan, where the multiplet satellite signals have *four per cent* of the intensity of the main $^{12}CH_3$ signal. This is because the molecule of dioxan contains four equivalent methylene groups, so that there is about a four per cent natural abundance of molecules in the solvent which contain one $^{13}CH_2$ group.

The pair of lines at 78 and 193 c/s (from T.M.S.) and the pair at 107 and 164 c/s constitute "spinning side-bands". The innermost lines are each about 28 c/s from the sharp signal at τ 7·75, and the outermost lines are each about 57 c/s from this. The lower of these two frequencies arises directly from the spinning of the sample tube in the spectrometer. The tube is almost certain to wobble very slightly, and the vibration will probably have a frequency the same as that of the rotation. The sample will then be cutting lines of magnetic force at this frequency, and so the absorption signal is modulated and side-bands appear, spaced on each side of the main signal at distances equal to this modulation frequency or to an integral multiple thereof—in the present case at about 28 and $2 \times 28 = 56$ c/s. Variation of the speed of rotation of the sample, alters the positions of the spinning side bands, and so these are easily distinguished from true spectral lines.

Spinning side-bands are then, effectively, the result of a slight mechanical defect. Obviously, sample tubes should be uniform in bore, perfectly cylindrical and a correct fit in the spectrometer probe. Even with full precautions, however, spinning side-bands become prominent when the spectrum contains very strong signals, such as arise from protic solvents. This is because the amplitude of the side-bands depends in part on that of the main signal.

Summary

Signal (c/s)	Position (ppm) δ	τ	Intensity (relative)	Multiplicity	J (c/s)	Assignment
135	2·25	7·75	3	singlet		CH_3–
396	6·60	3·40 (B)	1	doublet	16·5	$H_{(B)}$ (Ar)
444	7·40	2·60 (A)	1	doublet	16·5	$H_{(A)}$
432 to 456	7·20 to 7·60	2·80 to 2·40	5	complex		five protons on a benzene ring, C_6H_5

This substance is

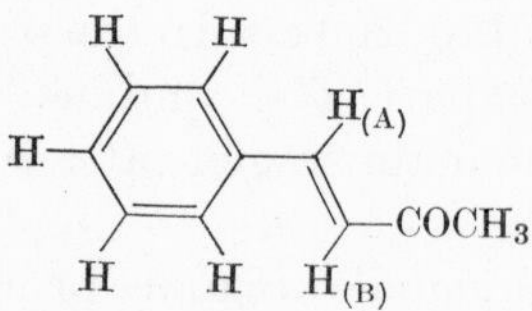

2. Compound C_7H_9N

Solvent: carbon tetrachloride
Sweep width: 500 c/s
Sweep offset: zero

The three bands in the spectrum presumably arise from three sets of protons in chemically distinct environments. The steps in the integral-trace over these bands are in the ratio 1 : 4 : 4, reading from left to right, and these numbers must represent the number of protons responsible for the separate signals because there are nine protons in all. The positions of the signals (see Tables 1 and 4, pp. 51 and 54) suggest that one olefinic proton and two pairs of methylene protons are present, so that the molecule comprises the following fragments:

$$\text{H}{>}\text{C}{=}\text{C}{<}\ (\tau\ 3{\cdot}40);\ CH_2,\ CH_2\ (\tau\ 7{\cdot}80);\ CH_2,\ CH_2\ (\tau\ 8{\cdot}32).$$

The splitting of the olefinic-proton signal into five lines suggests that it is coupled to *four* other protons. These could be the protons on two flanking methylene groups,

$$\cdot CH_2{-}CH{=}\overset{|}{C}{-}CH_2\cdot$$

Table 1 (p. 51) gives τ 8·05 for the protons of a methylene group adjacent to a double bond, and this value would move downfield by about 0·2 ppm if the group were also part of a ring, or even further if an additional deshielding substituent were present. The complex signal centred at τ 7·80 could thus be accounted for.

The remaining pair of methylene groups, evidently responsible for the signal centred at τ 8·32, could then be accommodated along with the $-CH_2CHC{:}CH_2-$ grouping as a cyclohexene ring. Completion of the molecule with a carbon and a nitrogen atom (to make C_7H_9N) gives cyclohexenyl cyanide. The cyanide substituent, attached to the double bond, provides a reason for the low-field positions of the several signals in the spectrum.

[It may be observed that there is no signal in the spectrum which immediately suggests the presence of an amino or imino group in the molecule. The fine structure of the low-field one-proton signal precludes its assignment to an :NH group. The occurrence of an $\cdot NH_2$ or :NH proton signal within the complex group of lines around τ 7·80 is conceivable, but insuperable structural difficulties are encountered, starting from this premise.]

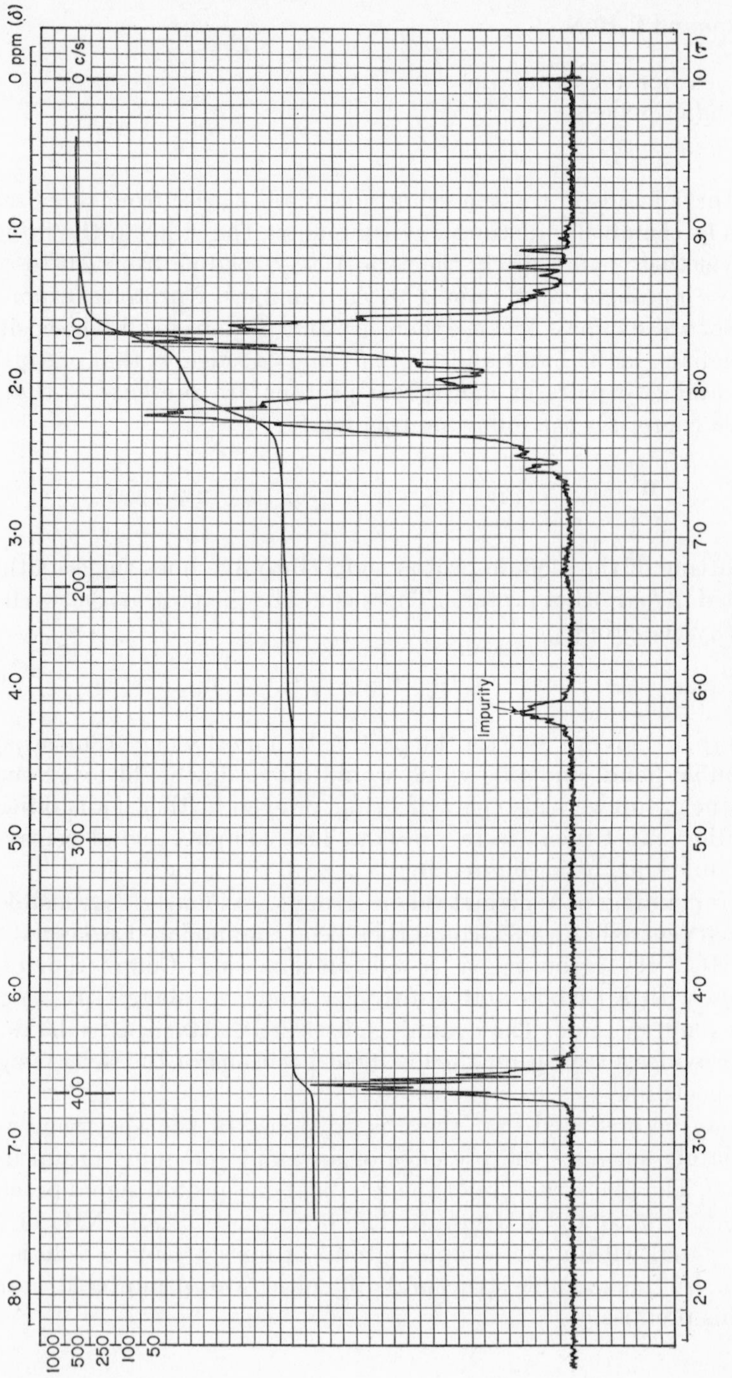

Spectrum of Compound 2.

Summary

Signal (c/s)	Position (ppm) δ	τ	Intensity (relative)	Multiplicity	Assignment
101	1·68	8·32	4	Complex	$-CH_2CH_2-$
132	2·20	7·80	4	Complex	Two CH_2 groups flanking a double bond.
396	6·60	3·40	1	Quintuplet	One olefinic proton coupled to the protons of two methylene groups.

This substance is

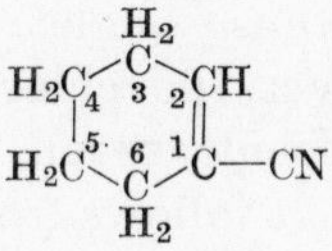

```
         H2
         C
  H2C4 /  3  \2CH
     |           ||
    5|    6     1||
  H2C \      /  C—CN
         C
         H2
```

3. Compound $C_5H_7N_3$

Solvent: deuterochloroform
Sweep width: 500 c/s
Sweep offset: 100 c/s*

The chart shows, in addition to the spectrum of the dissolved compound, the line from the internal reference, T.M.S., which is set at τ 10. There is also a line downfield at 442 c/s from T.M.S., which arises from chloroform ($CHCl_3$) present as an impurity in the deuterochloroform ($CDCl_3$) used as solvent. [The position of the line given by chloroform varies between about 435 and 445 c/s from T.M.S. (at 60 Mc/s), depending on the solute. It is of course essential to recognize lines which arise from the solvent, either intrinsically or from an impurity.]

The spectrum of the dissolved compound, then, shows two doublets, a broad band and a sharp intense singlet. The total rise in the integral trace (neglecting the rise over the chloroform line), divided by the number of protons in the molecule, seven, gives the size of a step in the integral trace for one proton. It follows from this information that each doublet signal arises from a separate proton, that the broad signal arises from two protons, and the intense line from three.

This last at τ 7·63 (intensity 3) must arise from the three equivalent protons of an isolated methyl group (otherwise there would be splitting of the signal). Its position suggests attachment to an aromatic ring. Table 1 (p. 51) gives τ 7·66 for the methyl group in toluene.

Protons attached to nitrogen frequently give broadened signals (Table 5, p. 56). This happens at certain rates of exchange and as a result of the nitrogen nucleus (with $I = 1$) undergoing changes in spin state. The broad signal of intensity 2 could therefore be assigned to the protons of an amino group: it's position at τ 4·43 suggests direct attachment to an aromatic ring.

The doublet signals have origin positions at τ 1·78 and 3·47, in the region characteristic of olefinic, aromatic and heteroaromatic protons (Table 4, p. 54). Each doublet shows the same splitting (which is absent from the other signals in the spectrum), and so the two protons responsible are coupled to one another. (The four lines constitute an *AB* quartet pattern.) The magnitude of the coupling, $J = 5$ c/s, suggests that the protons are *cis* about a double bond (Table 6, p. 57). [*Each* proton in the grouping

---

* The true position of any line is found by adding the "offset" figure to the chart position in c/s.

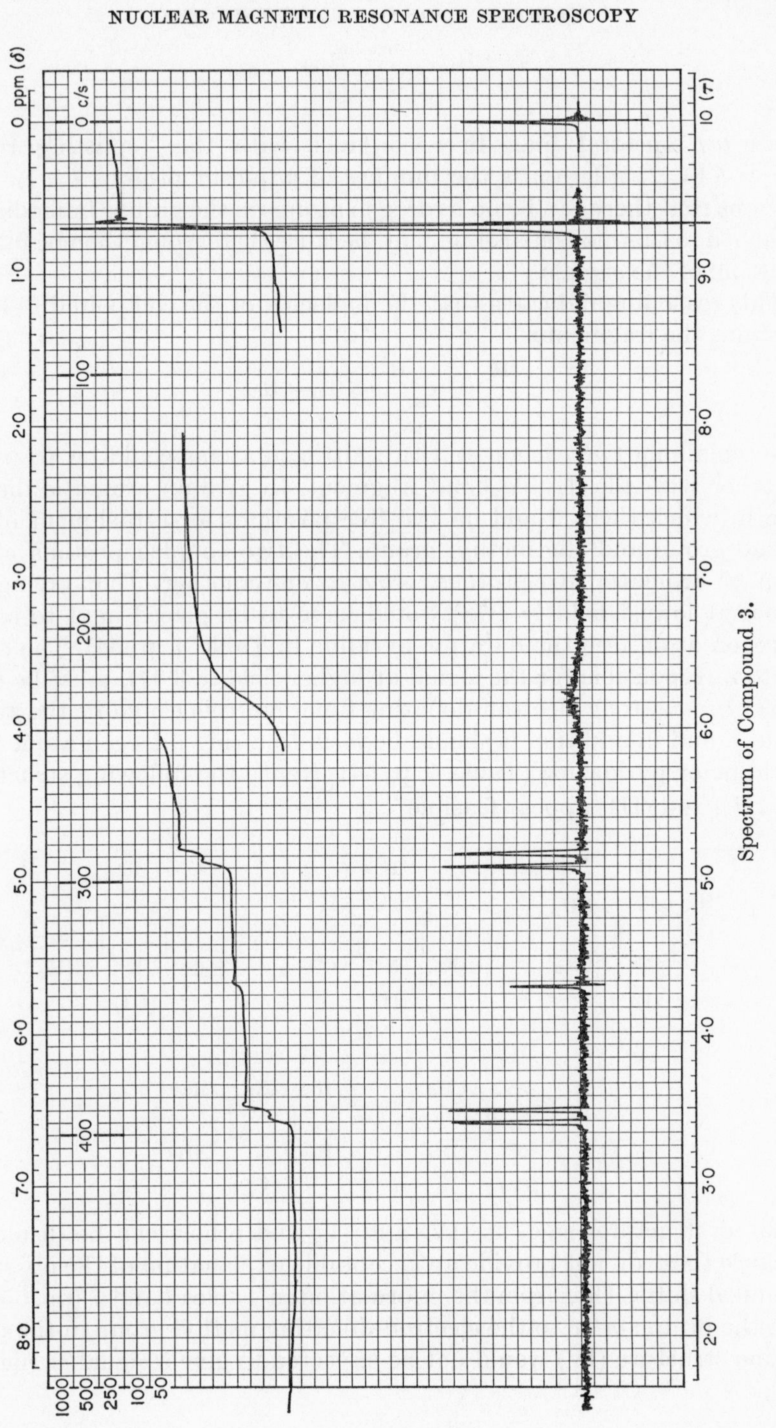

Spectrum of Compound 3.

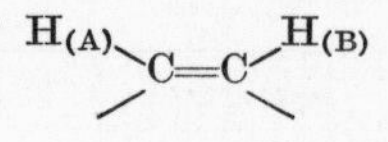

gives *two* spectral lines in accordance with the "splitting rule", $N = n + 1 = 2$, where n is the number of adjacent protons ($= 1$). It is obvious that there can be no hydrogen atoms on the groups immediately attached to this molecular fragment, because there would then be further splitting of the signals.]

This reasoning indicates that the molecule of the compound $C_5H_7N_3$ contains the fragments

$$\mathrm{H{-}C{=}C{-}H},\ -NH_2,\ -CH_3.$$

The remaining two carbon and two nitrogen atoms could be arranged, together with the double bond fragment, to give an aromatic diazine ring in which there would be two free positions for attachment of the amino group and the methyl group. The two coupled protons would then be aromatic ring-protons, *ortho* to one another. That giving the signal at lowest field (τ 1·78) would have to be placed next to a ring nitrogen atom (see the data for pyridine in Table 4, p. 54). The other proton, responsible for the signal at higher-field (τ 3·47), must be *ortho* or *para* to a strong electron-donating function (this would be the amino group), and in addition it should be *meta* to a ring nitrogen atom as in pyridine or pyridazine (Table 4, p. 54). Hence the following structures (I), (II), and (III) appear feasible:

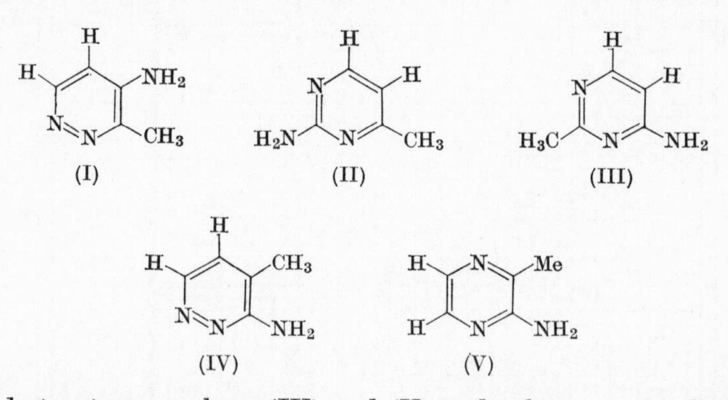

Related structures such as (IV) and (V) and others can be dismissed because these do not provide an acceptable explanation of the observed chemical shifts. Thus from the information in Tables 4 and 7 (pp. 55 and 58), the proton *ortho* to the electron-donating methyl group in the pyridazine structure (IV) would not be expected to give a signal as high as τ 3·5; up to ~τ 2·7 would appear more reasonable. The proton next to the ring nitrogen and *meta* to the amino group could hardly give a signal as high as τ 1·78: a value no higher than τ 1·2 would be expected. In structure (V), the signal from the proton *para* to the amino group would not be expected above τ 2·0.

A decision between the likely structures (I), (II) and (III) cannot be reached with complete certainty on the basis of the information given in the tables of τ values. It would be necessary to determine the spectrum of one (or more) of the possible compounds for comparison with the spectrum shown above. This kind of procedure is frequently essential in practice. When there is possible ambiguity, structural deductions made from chemical-shift data must be supported by spectral information from closely related model compounds.

A tentative decision, however, between the structures (I), (II) and (III), in favour of (II) can be reached by a consideration of the chemical shift of the methyl-group protons.

For a methyl group attached at the β- or γ-position in pyridine, the line position is τ 7·74 (Table 2, p. 52). In structure (I), the *ortho* amino group would raise this value by about 0·1 ppm to τ 7·84 (see footnote to Table 1, p. 51), a value appreciably higher than that found for the unknown compound. For a methyl attached at the α-position in pyridine, the chemical shift is τ 7·56 (Table 2, p. 52). This, in structure (II), would be raised by the *meta* amino group, perhaps to τ 7·66, a figure very close to the observed value. On the other hand, in structure (III), the methyl resonance would be expected at a lower value than τ 7·56, because of the *two* adjacent ring nitrogen atoms.

Summary

Signal (c/s)	Position (ppm) δ	τ	Intensity	Multiplicity	J (c/s)	Assignment
142	2·37	7·63	3	Singlet		CH_3—(Ār)
334	5·57	4·43	2	Broad		NH_2—(Ār)
392	6·53	3·47 (B)	1	Doublet	5	H(B)
493	8·22	1·78 (A)	1	Doublet	5	H(A) N=

This substance is

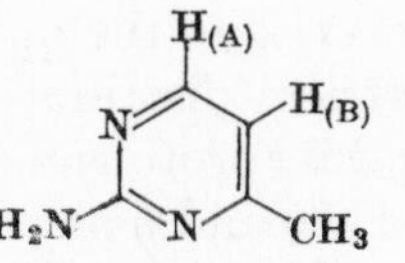

4. Compound $C_9H_{11}BrO$

Solvent: carbon tetrachloride
Sweep width: 500 c/s
Sweep offset: zero

The spectrum is of a solution in carbon tetrachloride, and all lines therefore arise from the solute.

The spectrum shows four groups of lines, which correspond to four sets of protons in chemically distinct environments. The steps of the integral trace are roughly in the ratio 5:2:2:2, reading from left to right, and, since the total number of protons is eleven, these figures must represent the numbers of protons involved in the multiplets severally. The 5-proton signal at about τ 3·0 lies in that region where phenyl group protons are to be found (τ 2·73) and Table 4 (p. 54) indicates that an electron-donating substituent, such as –OR could account for the observed shift to higher field, i.e. towards the right (a shift of up to 1 ppm may thus be observed). As a substituent on the phenyl group would destroy the equivalence of the remaining five benzenoid protons, so the complex pattern from about τ 3·3 to 2·6 ($\equiv$5 protons) can be accounted for. Reference to spectra of phenyl ethers will show how characteristic this pattern is.[2a]

The triplet centred near τ 6·0 is equivalent to two protons and so must arise from a methylene group. As the signal is split into *three* lines ($J = 5{\cdot}75$ c/s), the two equivalent protons of this methylene group must be coupled to another *pair* of equivalent protons. For similar reasons, the triplet centred near τ 6·5 ($J = 6{\cdot}5$ c/s) can be assigned to a second methylene group also coupled in turn to another methylene group. As the coupling constants for the above two groups (at τ 6·0 and 6·5) are not the same, it is concluded that these methylenes cannot be coupled to one another. Each must be coupled, with a slightly different J value, to a third methylene group. This is presumably responsible for the quintet signal centred at τ 7·80. An additional pointer to such a conclusion is the distortion in the intensities of the lines in the two triplets, which depart from the ideal ratio 1:2:1. Of the two outer lines of each triplet it is the higher field line (i.e. right-hand line) which is the more intense—an indication that the group to which both of the responsible protons are coupled, is to be found at a higher field. This group can only be the methylene group which gives rise to the quintet signal, and this in turn is distorted from a symmetrical distribution of relative intensities in such a way as to indicate that its protons too are coupled to others which lie at lower field (in fact at τ 6·0 and 6·5).

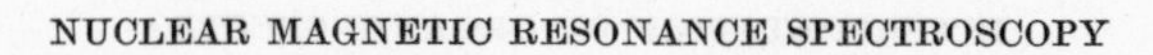

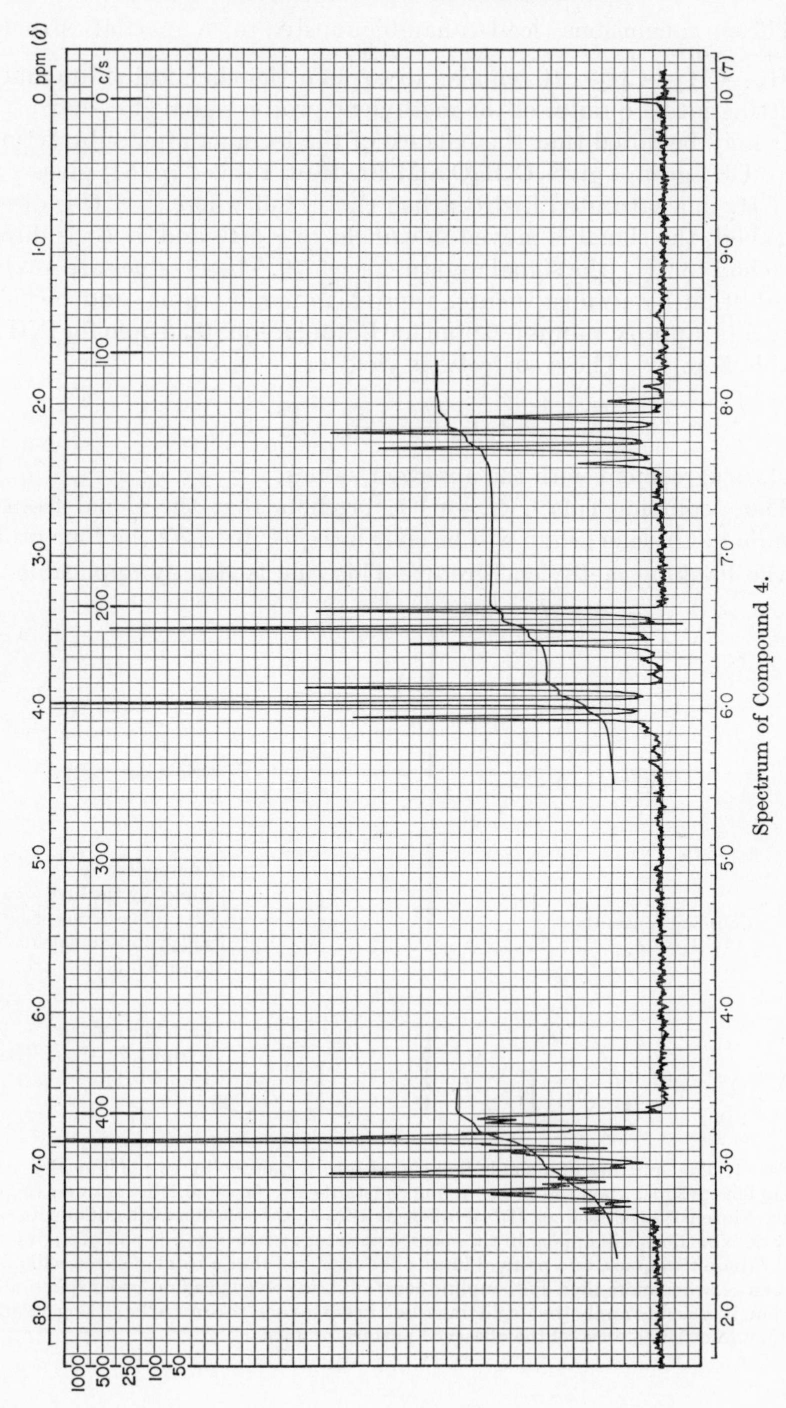

Spectrum of Compound 4.

These conclusions lead unambiguously to a partial structure, $\overset{\tau=6\cdot0}{-CH_2}-\overset{7\cdot80}{CH_2}-\overset{6\cdot5}{CH_2}-$. It remains to explain the chemical shifts and the splitting into a quintet of the methylene proton signal at τ 7·80.

It may be noted that the balance of the formula after subtraction of three CH_2 groups gives C_6H_5O and Br. Now Table 1 (p. 51) gives τ 8·75 for CH_2 in a saturated hydrocarbon chain, but where C_6H_5O is directly attached, the signal is moved downfield to τ 6·10 and, if Br is directly attached to CH_2, the signal appears at τ 6·70. These values are modified in addition by γ-substituents, approx. minus 0·1 ppm, and by β-substituents, approx. minus 0·6 ppm for Br and minus 0·35 ppm for C_6H_5O–, (Table 2, p. 52). These corrections yield

$$C_6H_5O-\overset{\tau=6\cdot0}{CH_2}-\overset{7\cdot8}{CH_2}-\overset{6\cdot6}{CH_2}-Br$$

in close agreement with the observed values.

The "splitting rule", $N = n+1$, predicts that the signal from the middle methylene group will be split into *five* lines, by the four protons of the flanking methylene groups. This rule is strictly applicable only

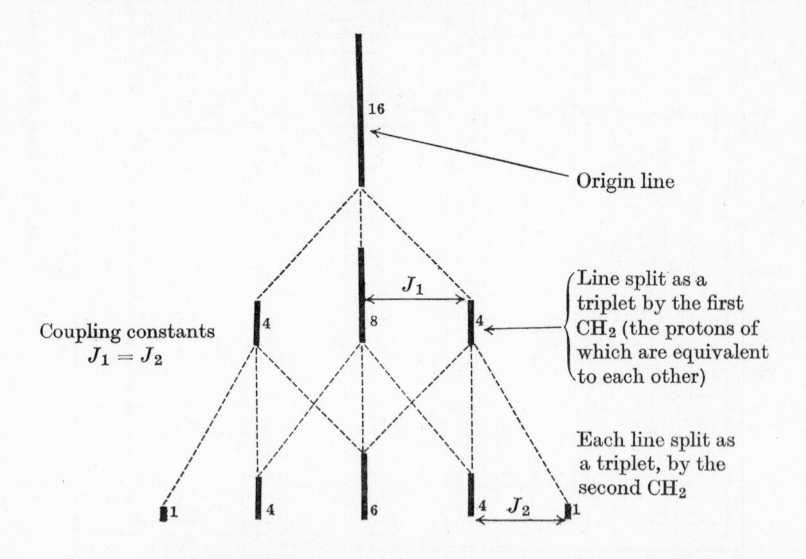

FIG. 1. The splitting of a single-line signal into a quintet by two chemically distinct methylene groups, when the two coupling constants are the same. The numbers represent relative intensities of the lines. If J_2 is different from J_1, then the lines will not be superposed as here. Nine lines will appear in the spectrum, except when J_2 is one half of J_1 or is twice J_1; in these two cases, two pairs of lines will overlap so that a total of seven will appear. Therefore, when more than one coupling constant is involved in a first-order spin multiplet system, it is best to apply the "splitting rule" in stages and to draw a "splitting diagram", as here, in order to unravel the observed pattern of lines.

for a single coupling constant. In the present case however the *two* couplings are sufficiently similar, numerically, for the rule to give a correct result (cf. Fig. 1). Ideally the relative intensities of the five lines should be 1:4:6:4:1, but there is a distortion from this, the low-field lines of the quintet being increased in intensity at the expense of the high-field lines. The reason, already mentioned, is that the protons responsible for the quintet signal are coupled to the protons of two methylene groups at *lower-field* (i.e. to the left), and there is not a great separation between the signals. If there were a large separation, then the lines of the quintet (and of the triplets) would have intensities closely approximating to the ideal.

Summary

Signal (c/s)	Position (ppm) δ	τ	Intensity	Multiplicity	J (c/s)	Assignment
132	2·20	7·80	2	Quintet		CH_2 flanked by two more CH_2 groups
209	3·48	6·52	2	Triplet	6·5	CH_2Br next to CH_2.
238	3·97	6·03	2	Triplet	5·75	$CH_2\cdot O\cdot Ph$ next to CH_2.
400 to 443	6·67 to 7·36	3·33 to 2·62	5	Complex		Five protons on aromatic ring

This substance is

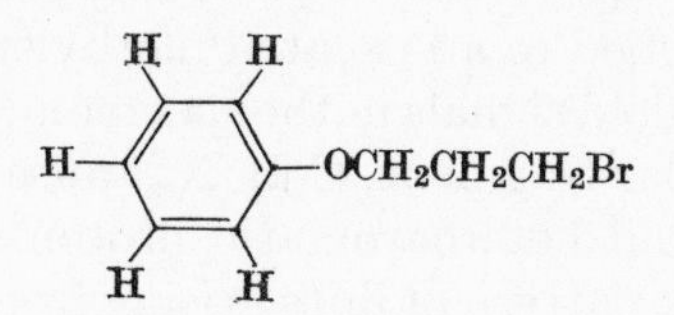

5. Compound $C_9H_{11}NO_2$

Solvent: carbon tetrachloride
Sweep width: 500 c/s
Sweep offset: 30 c/s*

A striking feature of the spectrum is the triplet and quartet signals centred at τ 8·78 and 5·9 respectively, which have a common line separation of 7 c/s and so may constitute the components of a spin-spin multiplet system. The positions of the multiplets and the pattern of seven lines are recognizable after very little practice as the multiplet signal from an ethoxyl group, an A_2X_3 spin-spin system. It is important in making this assignment to note that the ratio of the total intensity of the quartet and of the triplet signal, as given by the integral trace, is 2 : 3, and that the intensities of the individual lines within each multiplet are roughly in the ratios 1 : 3 : 3 : 1 and 1 : 2 : 1 respectively. A detailed explanation of this splitting pattern was given on p. 3. The chemical shifts suggest that the ethoxyl group is part of an ester function. Thus, Table 1 (p. 51) gives τ 5·85 for the shift of a methylene group in the environment $-CH_2-OCOR$, and Table 2 (p. 52) indicates that the terminal methyl group would have a shift of τ $9{\cdot}10 - 0{\cdot}27 = 8{\cdot}83$ (the allowance being for the β-oxygen atom). The assignment

$$\overset{\tau = 8\cdot78}{CH_3}-\overset{5\cdot9}{CH_2}-O-CO-$$

therefore seems certain.

The singlet at τ 6·25 arises from *two* protons, as indicated by the integral trace, and can be assigned to an isolated methylene group, $-CH_2-$.

The remaining low-field signals in the spectrum, which must arise from the remaining molecular fragment, C_5H_4N, appear in the region characteristic of aromatic and heteroaromatic protons. Between τ 2·25 and 3·08 there is a complex pattern of lines arising from three protons, which are evidently coupled to one another. At τ 1·52, there is a single-proton signal. This is a doublet ($J \approx 5$ c/s) which shows signs of fine splitting, so that the proton must be coupled strongly to one other proton and weakly to others. These observations suggest a pyridine residue,

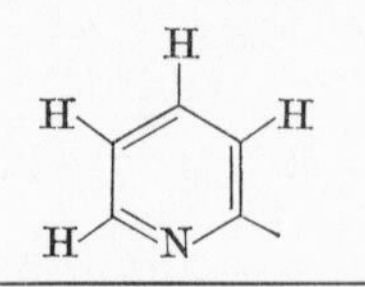

* See footnote p. 11.

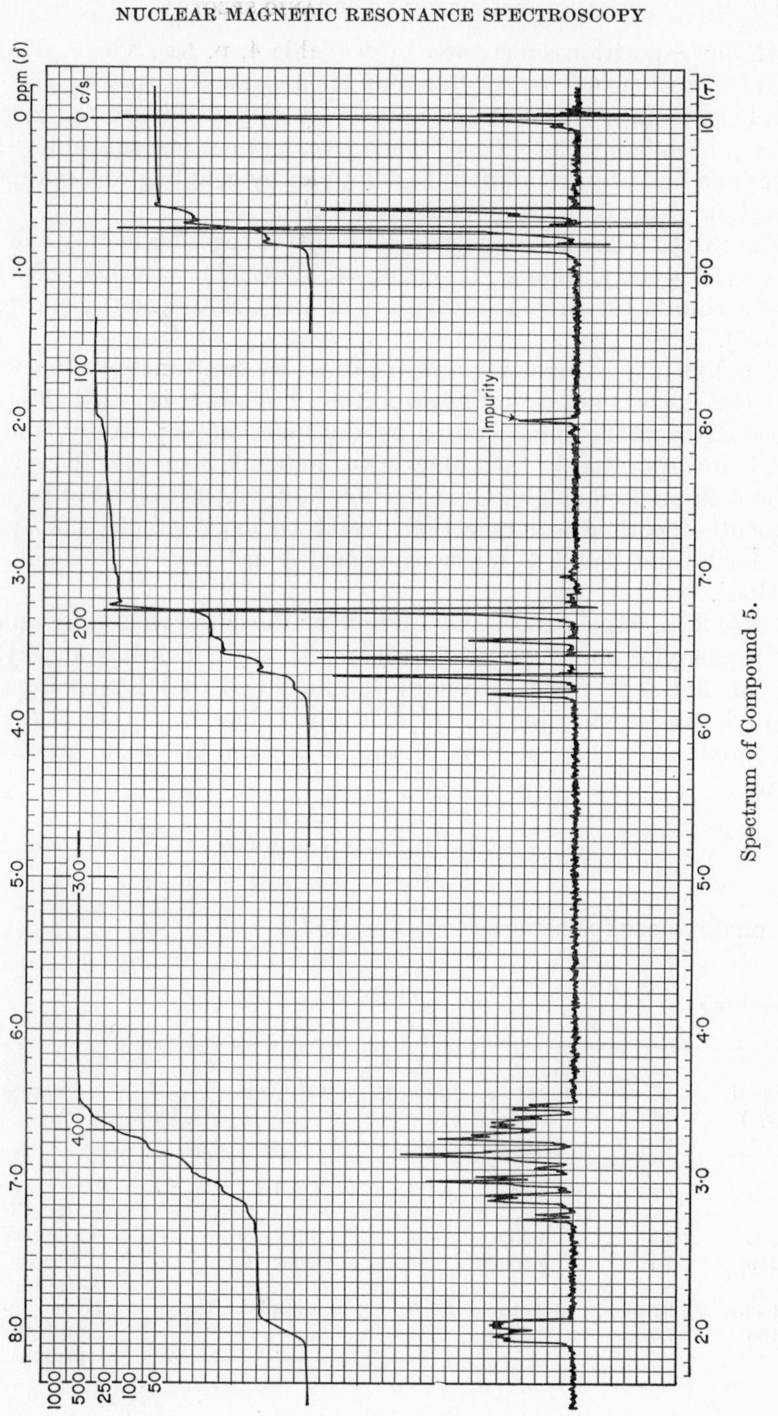

Spectrum of Compound 5.

(with one α-position substituted) (see Table 4, p. 54). The β, β' and γ protons have insufficiently different chemical-shifts to give rise to a simple splitting pattern and are evidently responsible for the complex three-proton band. Only the α-proton gives a separate signal at low-field, split (to a first approximation) as a doublet by coupling to the adjacent β-proton (Table 6, p. 57, gives $J_{23} = 5{\cdot}5$ c/s).

The three molecular fragments can be fitted together in only one way, as ethyl 2-pyridylacetate. It remains to determine whether this structure accounts for the chemical shift of the isolated methylene group (τ 6·25).

The difficulty of assessing a shift caused by two functions together is that the effect is not accurately additive. Usually it is a little less than expected from the shifts caused by the functions separately, and this fact is incorporated in the values gives in Table 3 (p. 54): these values are not to be used singly. Occasionally, however, the effect of two substituents together is larger than could be expected, so that values obtained from Table 3 must be regarded as only very rough predictions.

Table 3 (p. 54) gives 0·7 and 1·3 ppm as the shifts caused by an ester and a phenyl group respectively when both are attached to methylene. The phenyl shift could reasonably be increased to 1·4 ppm for an α-pyridyl substituent, judging from the τ values in Table 2 (p. 52). The predicted chemical shift of the methylene group in question is then

$$\tau\ 8{\cdot}75 - (0{\cdot}7 + 1{\cdot}4) = 6{\cdot}65,$$

reasonably near to the observed value, 6·25.

Summary

Signal (c/s)	Position (ppm) δ	τ	Intensity (relative)	Multiplicity	J (c/s)	Assignment
73	1·22	8·78	3	Triplet	7	$CH_3\cdot$ (of CH_3CH_2O)
225	3·75	6·25	2	Singlet		$\cdot CH_2 \cdot$ (isolated)
246	4·10	5·90	2	Quartet	7	$\cdot CH_2 \cdot$ (of CH_3CH_2OCO)
415 to 465	6·92 to 7·75	3·08 to 2·25	3	Complex		Three coupled aromatic protons

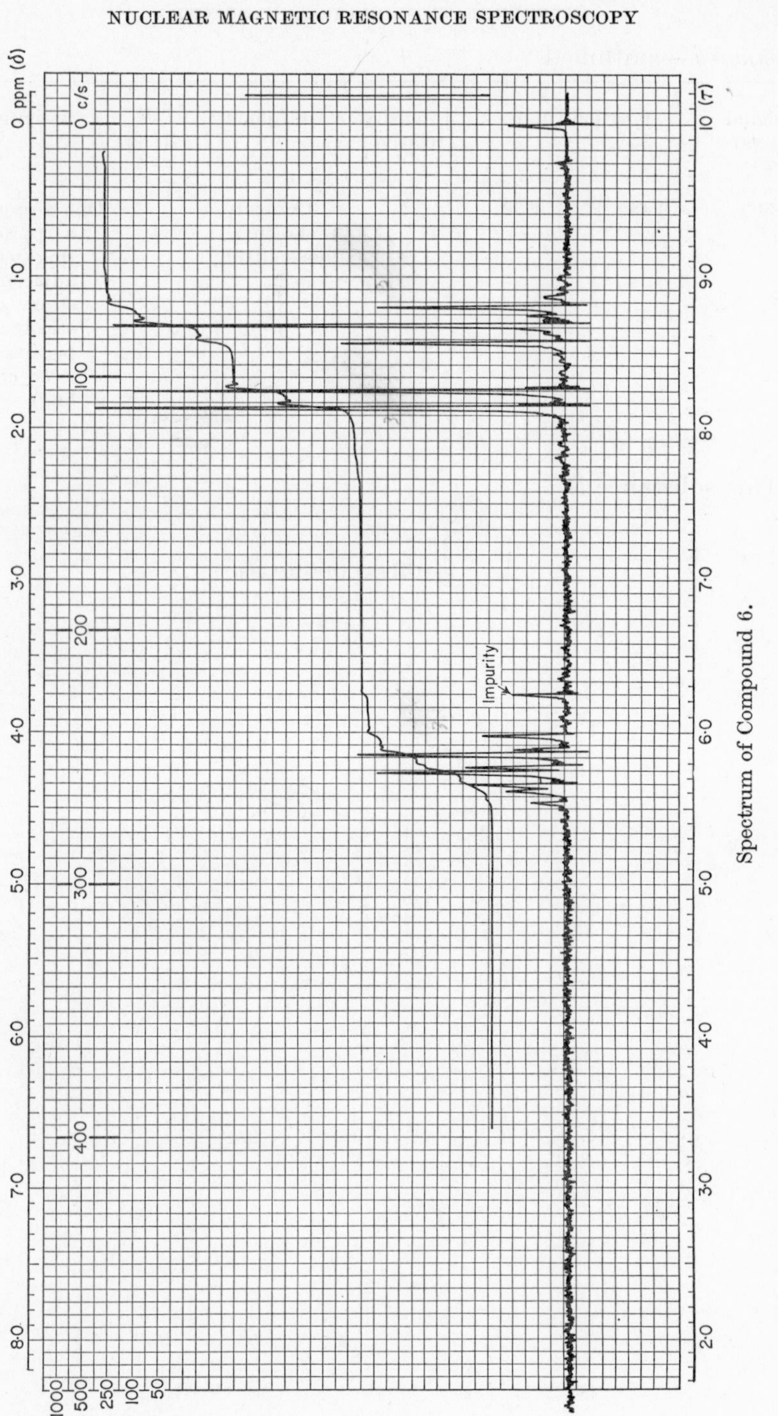

Spectrum of Compound 6.

Summary—continued

Signal (c/s)	Position (ppm) δ	τ	Intensity (relative)	Multiplicity	J (c/s)	Assignment
509	8·48	1·52	1	Doublet, showing fine splitting	~5	One α-proton on a pyridine ring (with the adjacent position unsubstituted, because of the *ortho* coupling)

This substance is

(2-pyridyl)—CH_2—$COOCH_2CH_3$

6. Compound $C_5H_9BrO_2$

Solvent: carbon tetrachloride
Sweep width: 500 c/s
Sweep offset: zero

There appear to be three groups of signals in the spectrum; a group of eight lines, a doublet and a triplet. Each group has the same intensity, as indicated by the integral trace. Because there are nine protons in all, the groups of lines must arise from three protons each.

The triplet signal at τ 8·69 and the doublet at 8·2 have the same splitting, $J = 7$ c/s, but cannot be related spin-multiplets because both are *three*-proton signals and the "splitting rule" would require both to be split as *quartets*.

The 1:2:1 triplet (at τ 8·69) must arise from the protons of a methyl group coupled equally ($J = 7$ c/s) to *two* other protons. This leads to a structural fragment CH_3CH_2–. The signal from the methylene, yet to be located, would appear as a 1:3:3:1 quartet of lines, unless there were further coupling.

The doublet signal (at τ 8·2) must arise from the protons of a methyl group coupled to *one* other proton, so that a second structural fragment is CH_3CH–. The signal from the single proton would be split as a 1:3:3:1 *quartet* of lines by the *three* protons of the adjacent methyl, provided there was no further coupling.

The signals from the single proton and from the methylene group must give rise to the low-field group of eight lines, which then consists of two quartets. The separate quartets are easily identified. Each has $J = 7$ c/s and the lines of one are twice as intense as the corresponding lines of the other. The two-proton quartet has an origin position at τ 5·80 and that of the one-proton quartet is τ 5·72.

The spectrum therefore really contains four groups of signals which constitute two spin-spin multiplet systems. These are shown separately (in ideal form) below.

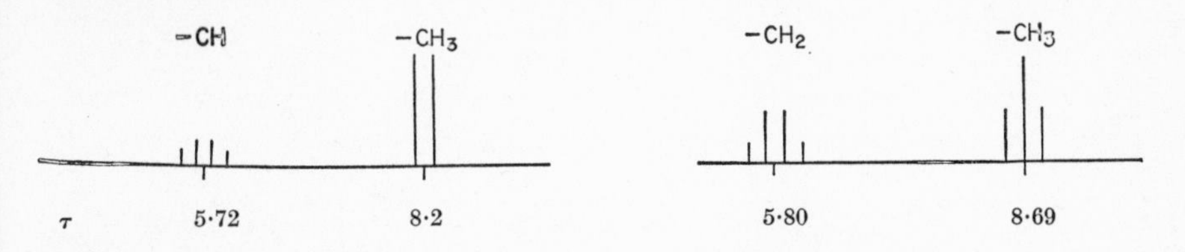

The chemical shift of the methylene group indicates that it is attached to oxygen in an ester function: Table 1 (p. 51) gives τ 5·85 for $-CH_2OCOR$.

Hence the compound must be an ethyl ester, CH_3CH_2OCO–. A predicted chemical shift for this methyl group β- to oxygen (from Table 2, p. 52), is τ 9·10 − 0·27 = 8·73, in close agreement with the observed value, 8·69.

It is only possible to complete the structure, with the other fragments $CH_3CH\lessdot$, and Br- as ethyl α-bromopropionate, $CH_3CH_2OCOCHBrCH_3$. In support, we find from Table 2 (p. 52), a predicted chemical shift for the methyl next to CHBr of τ 9·10 − 0·81 = 8·29. A predicted value for the CH proton will be less accurate and Table 3 (p. 54) indicates τ 8·75 − (1·9 + 0·7) = 6·15, but this is not unduly far from the value observed.

Summary

Signal (c/s)	Position (ppm) δ	τ	Intensity	Multiplicity	J (c/s)	Assignment
78·5	1·31	8·69	3	Triplet	7	CH_3· of CH_3CH_2O
108	1·80	8·20	3	Doublet	7	CH_3· of $CH_3CH(CO)(Br)$
252	4·20	5·80	3 (252 and 257 together)	Quartet	7	CH_2 of CH_3CH_2OCO
257	4·28	5·72		Quartet	7	CH of $CH_3CH(CO)(Br)$

This substance is

$$CH_3CH_2OCOCHBrCH_3$$

7. Compound $C_8H_{12}O$

Solvents: *A*, in CCl_4; *B*, after this solution had been shaken with acid; *C*, in pyridine
Sweep width: 500 c/s
Sweep offset: zero

The steps in the integral trace over the two lines and band in the spectrum *A* are in the ratio 1:1:10. These numbers must represent the numbers of protons involved because there are twelve in all.

The one-proton signal at τ 6·51* (spectrum *A*) is somewhat broader than a normal signal. It is sharpened and shifted a little by a trace of acid (as in spectrum *B*) and is shifted considerably in pyridine (spectrum *C*). So this signal can be assigned to a hydroxyl group (cf. Table 5, p. 56).

Table 1 (p. 51) gives τ 7·61 as the chemical shift for a CH group attached to –CHO, but this cannot be a correct assignment for the sharp one-proton signal at τ 7·63 since the signal would be split by coupling to the adjacent aldehyde proton. Moreover, the signal is moved to lower-field by pyridine (spectrum *C*), so that it evidently arises from a proton that can associate with the solvent. The ethynyl group, –C≡C—H, for which Table 4 (p. 54) gives τ 7·6 when it is attached to a carbinol group, is therefore a likely possibility. Again, because the signal is a singlet, there cannot be a proton attached to the carbon atom next to the ethynyl group, and so the molecule must contain the fully substituted fragment –C—C≡C—H.

To this fragment must be attached the hydroxyl group and the C_5H_{10} residue. The latter, responsible for the broad signal (of intensity ten) around τ 8·4, presumably comprises five methylene groups. So the compound is evidently ethynylcyclohexanol.

It remains to account for the considerable broadening of the methylene signal—which in cyclohexane appears as a sharp line near τ 8·55, as indicated by Table 1 (p. 51). In the present case the substituents deshield the two flanking methylene groups perhaps by 0·15 ppm (Table 2, p. 52), so that a predicted chemical shift for these two groups is τ 8·40. In all probability, the ethynyl group will be equatorial so that the ring will have a fixed conformation. This will result in non-equivalent axial and equatorial protons, signals from the latter being shifted by about 0·2 ppm to lower field.[1a] A result of non-equivalence among the ring protons, (whatever the cause) is that the couplings between them will give rise to observable splitting. The spread of the resonance signals and

* Note that the chart paper used for this spectrum shows only a δ scale.

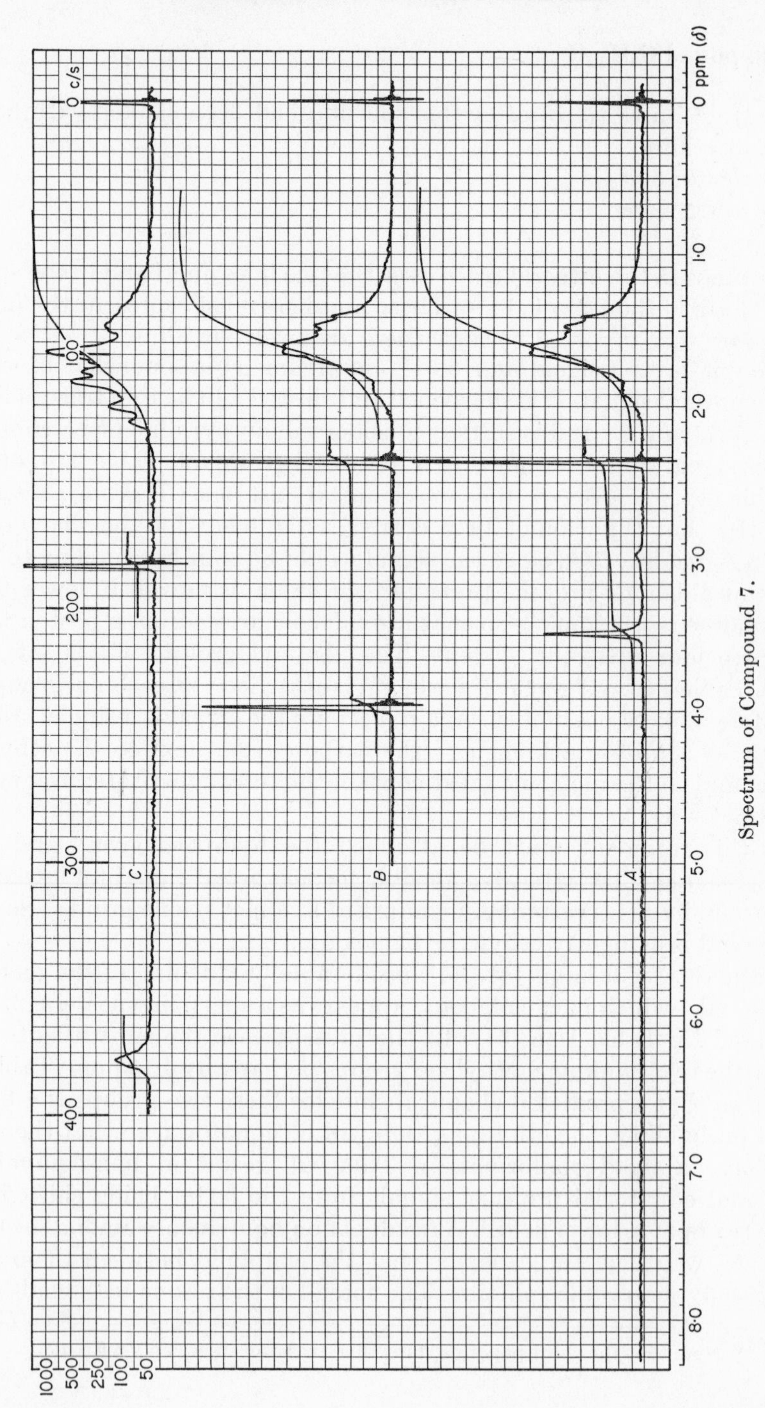

Spectrum of Compound 7.

particularly the considerable splitting of the individual signals by geminal and vicinal couplings (cf. Table 6, p. 57, for dependence of J upon bond angles) give rise to a very broad band.

Summary (Spectrum *A*)

Signal (c/s)	Position (ppm) δ	τ	Intensity	Multiplicity	Assignment
55 to 130	0·92 to 2·17	~8·4	10	Broad band	·$[CH_2]_5$·
142	2·37	7·63	1	Singlet	>C—C≡C—H
209·5	3·49	6·51	1	Broadened singlet	–OH

This substance is

$$\text{cyclohexane ring: } H_2C\langle C(H_2)\rangle C(OH)(C\equiv C{-}H)\text{, ring } -CH_2-C(H_2)-CH_2-$$

8. A Lactone $C_{10}H_{12}O_4$

Solvent: carbon tetrachloride
Sweep width: 500 c/s
Sweep offset: zero

A feature at once recognizable is the A_2X_3 pattern of lines from an ethyl ester group, $.COOCH_2CH_3$ (see Spectra 5 and 6, and p. 3). The 1:3:3:1 *quartet* (intensity two) at τ 5·70* arises from the methylene protons, which are coupled ($J = 7$ c/s) to the *three* protons of the methyl group. The latter gives the signal (intensity three) at τ 8·64 which is split as a 1:2:1 *triplet* by the coupling with the *two* methylene protons. The chemical shift of the methylene protons (τ 5·70) indicates that the ester function is attached to an aromatic, rather than aliphatic, residue (predicted values are τ 5·77 and 5·85, respectively, from Table 1, p. 51). In agreement, the chemical shift of the terminal methyl group (τ 8·64) is lower than that predicted from Table 2 (p. 52) (τ 9·10 − 0·37 = 8·73) because of the deshielding influence additional to that of the .OCO. function.

The three-proton singlet at τ 7·63 can be assigned to an isolated methyl group attached to an aromatic-type ring (Table 1, p. 51).

The doublet at τ 7·82 has an intensity of three (from the integral trace), and so must be assigned to a methyl group Because all the oxygen atoms in the compound have already been accounted for (a lactone function—given—and an ethyl ester group), the chemical shift can only be taken to mean attachment to an aromatic-type ring The value (τ 7·82) is higher than that for toluene (τ 7·66), but not so high as that for a methyl attached to a polyene chain (8·05) (see Table 1, p. 51). The small doublet splitting ($J = 1$ c/s) implies weak coupling to *one* other proton, which must be that responsible for the signal in the olefinic region of the spectrum, at τ 4·13, which shows signs of being split as a quartet The value of the coupling constant (1 c/s) signifies that the two coupled groups are separated by a double bond (see Table 6, p. 57).

There appear then to be the following fragments in the molecule:

$$-COOCH_2CH_3,\ -CH_3,\ -CH{=}C(CH_3)\diagdown \quad \text{and} \quad \overline{-CO-O-} \text{ lactone}$$

There remain two carbon atoms for completion of the structure. These two, the lactone function and the olefinic group, would together provide a six-membered ring, which, if it were fully unsaturated, i.e. an α-pyrone

---

* Note that the chart paper used for this spectrum shows only a δ scale (cf. those of compounds 7 and 11).

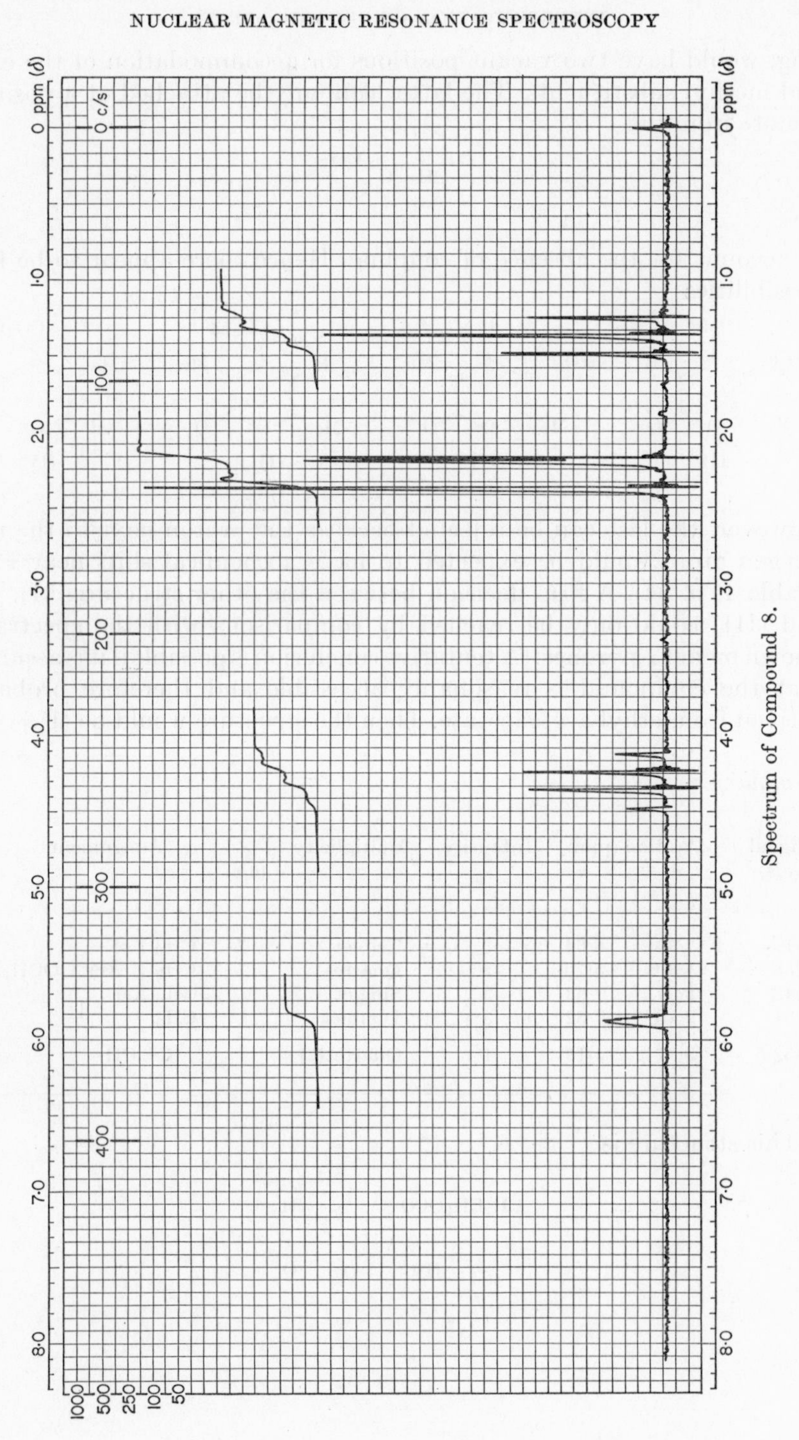

Spectrum of Compound 8.

ring, would have two vacant positions for accommodation of the ester and methyl substituents. The latter can only be attached at a position remote from the

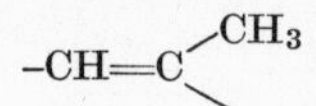

to account for the absence of coupling. Hence there appear to be four possibilities:

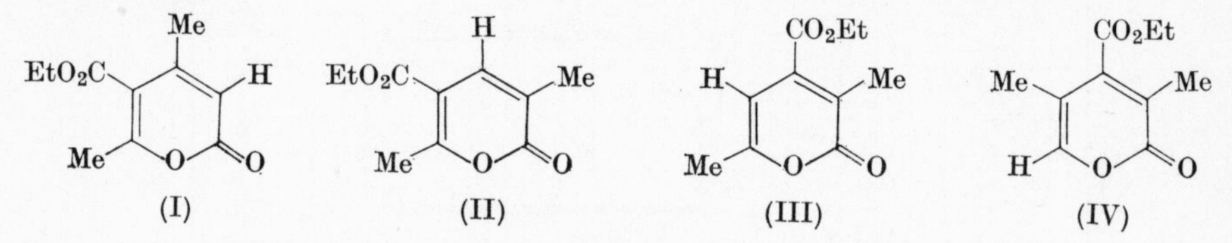

However, the last can be rejected because the proton next to the ring oxygen atom would be expected to have a chemical shift near τ 3·2 (Table 4, p. 54). A firm decision between the other structures (I), (II) and (III) could only be reached by comparisons with the spectra of known model α-pyrones. [A tentative conclusion is possible if it is assumed that the compound is reasonably accessible and therefore probably derived from ethyl acetoacetate. Then the structure would be (I).]

Summary

Signal (c/s)	Position (ppm) δ	τ	Intensity	Multiplicity	J (c/s)	Assignment
81·5	1·36	8·64	3	Triplet	7	CH_3 } of
258	4·30	5·70	2	Quartet	7	CH_2 } $(\bar{A}r)COOCH_2CH_3$
142	2·37	7·63	3	Singlet		$CH_3(\bar{A}r)$
131	2·18	7·82	3	Doublet	1	CH_3
352·5	5·87	4·13	1	Quartet(?)	1	$\rangle C{=}CH-$

This structure is

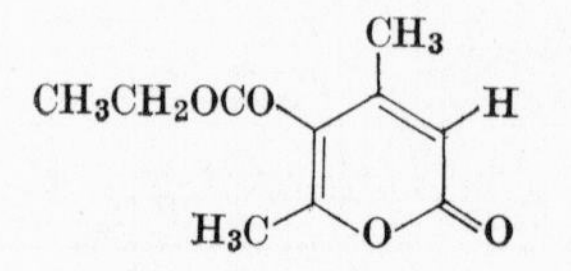

9. Compound $C_{10}H_{12}N_4O_4$

Solvent: deuterochloroform
Sweep width: 500 c/s
Sweep offset: zero; 200 c/s for *A*

It must be remembered that a line near τ 2·7 arises from chloroform in the solvent (cf. p. 12).

Calibration of the integral trace, by dividing the total rise by the number of protons (twelve), shows that the strong lines at 71 and 78 c/s from T.M.S., which are of unequal intensity, *together* constitute the signal from six protons. The only explanation is that this is a signal (τ 8·76) from the six equivalent protons of the *geminal* methyls of an isopropyl group. The signal is split as a doublet (J = 7 c/s) by coupling to the CH proton, which in turn will give a signal to lower-field, split into seven lines (unless there is further coupling, when the complexity will be greater). The broadened one-proton signal centred at τ 7·3 which must arise from the proton in question is not a simple septuplet, so there appears to be further coupling to one or more other protons. On the low-field side of the chloroform line, in the region where aromatic and olefinic protons give signals, there is a doublet of intensity one at τ 2·40. With J = 5 c/s, this bears no obvious relation to the remaining signals none of which shows a 5 c/s splitting. The relative intensities of the lines of the doublet—that at higher-field being slightly more intense—indicates that the single proton to which there is coupling gives a signal to higher field in the spectrum. This proton must then be that of the isopropyl group. Therefore the molecule contains an isobutylidene fragment

$$=CHCH\begin{matrix}CH_3\\CH_3\end{matrix}.$$

The next eight lines in the spectrum, from 470 to 550 c/s from T.M.S., in the low-field aromatic region, comprise (as the integral trace shows) three one-proton signals. The splittings indicate that the protons are coupled in two pairs. There is a low-field doublet at τ 0·88 with J = 2·5 c/s, and a higher-field doublet with J = 10 c/s, which has an origin at τ 2·07. In between, at τ 1·67, there is a one-proton signal which appears as a double-doublet, showing both of the previous splittings, 10 c/s and 2·5 c/s, so that the proton responsible must be coupled to both of the other two protons. The magnitudes of the two couplings correspond to those between protons disposed *ortho* and *meta*, respectively, on a benzene ring (see Table 6, p. 57). A pyridine ring is not possible. Hence there must

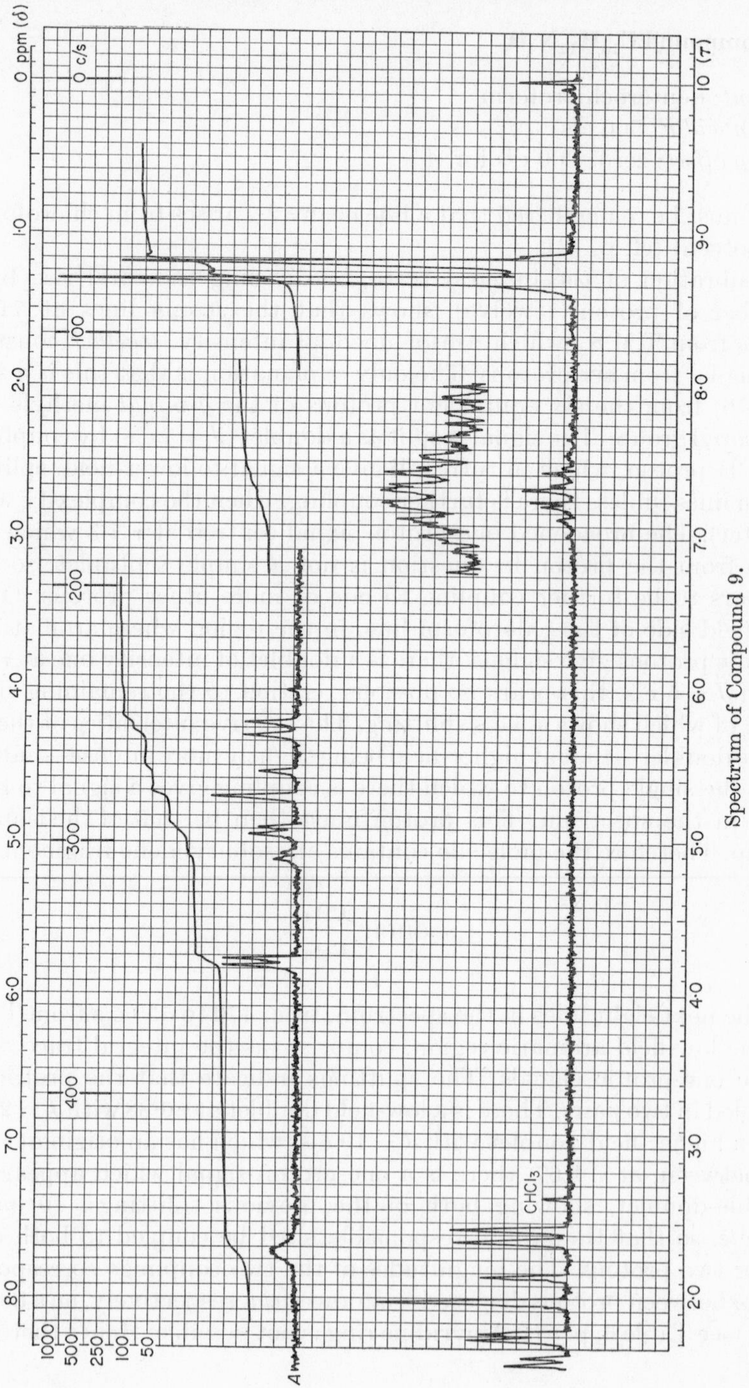

Spectrum of Compound 9.

be present a trisubstituted benzenoid fragment with the protons arranged 1,2,4:

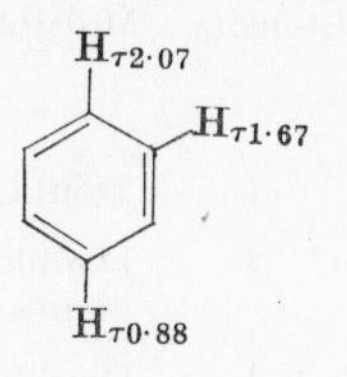

(evidently the *para* coupling is zero). We then have to account for the very low-field positions of the three protons' resonance lines. There must be two strong electron-withdrawing substituents to account for a shift from the normal benzene resonance at τ 2·73 to a value as low as τ 0·88: one such substituent cannot be expected to produce a shift of more than 1 ppm to lower field (see Table 7, p. 58). The lowest-field benzenoid proton could, then, be flanked by two *nitro* groups (suggested in part by the molecular formula), as follows:

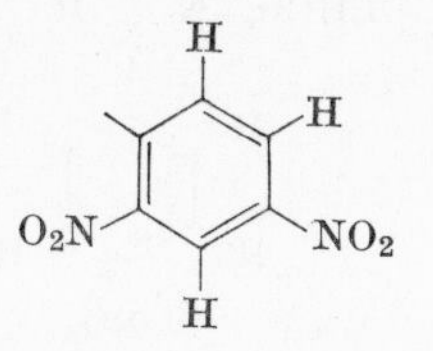

and the chemical shifts of the other two protons would not be in conflict with this.

Subtraction of dinitrophenyl and isobutylidene fragments from the molecular formula leaves a residue, HN_2. There is a broadened one-proton signal at τ $-1{\cdot}03$ (662 c/s from T.M.S.) and this could arise from the proton on an NH group, provided it was strongly deshielded, e.g. by chelation to oxygen, as would occur if there were an adjacent nitro group:

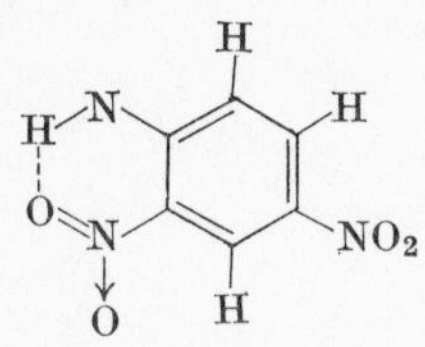

In agreement, the highest field benzenoid proton is then found to be that *ortho* to the NH group. The last fragment, N, and an isobutylidene residue can only be joined together to give the molecule of isobutanal 2,4-dinitrophenylhydrazone.

Summary

Signal (c/s)	Position (ppm) δ	Position (ppm) τ	Intensity	Multiplicity	J (c/s)	Assignment
74·5	1·24	8·76	6	Doublet	7	H_3C CH_3
135 to 185	~2·7	~7·3	1	Double septuplet(?)		CH
456	7·60	2·40	1	Doublet	5	=CH
476	7·93	2·07 (A)	1	Doublet	10	H(A)
500	8·33	1·67 (B)	1	Double-doublet	10 2·5	(B)H
547·5	9·12	0·88 (C)	1	Doublet	2·5	O_2N NO_2 H(C)
662	11·03	−1·03	1	Broadened		—NH...O

The structure is

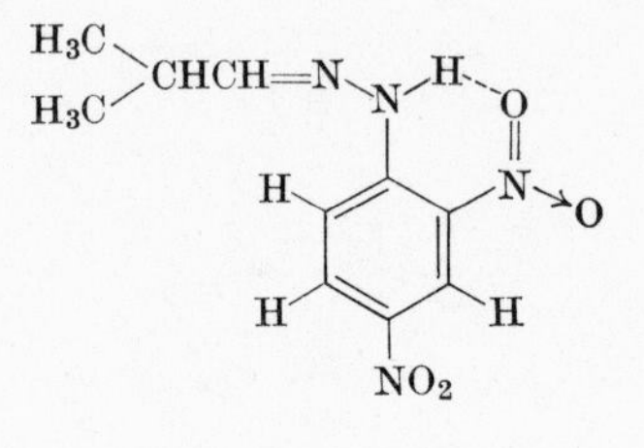

10. Compound $C_{11}H_{13}NO$

Solvent: deuterochloroform
Sweep width: 500 c/s
Sweep offset: 50 c/s

The sharp line in the middle of the group of signals in the aromatic region of the spectrum is very probably the chloroform line (at 437 c/s from T.M.S.). So the integral trace here must be corrected for an increment from this "impurity". A reasonable allowance would be half of the rise over the central peak which appears to comprise the chloroform line and one other. Division of the adjusted total rise in the integral trace by the number of protons (thirteen) then gives an accurate calibration.

The group of lines around τ 2·7* is then seen to comprise five protons: these are evidently coupled together and so could be those of a phenyl group, C_6H_5. In that case, the one-proton signal at τ 1·58 would not arise from a phenolic hydroxyl (see Table 5, p. 54); but it could be assigned to an amidic proton, ·CONH·, particularly as the signal is rather broad. The spread of the aromatic proton signals about the position characteristic of benzene (τ 2·73) suggests that the substituent is neither strongly electron-withdrawing nor strongly electron-attracting: the amide group would therefore be acceptable. Between τ 7·6 and 7·0 there is a *symmetrical* pattern of lines arising from *four* protons: these must constitute an A_2B_2 spin-coupled system.[1a] So we have the grouping X—CH_2—CH_2—Y present, where X and Y are almost identical in their shielding characteristics and have yet to be identified. The three-proton singlet to higher-field must arise from an isolated methyl group. Its chemical shift, τ 7·85, could be accommodated by its attachment to a ketonic carbonyl, as CH_3·CO–, for which Table 1 (p. 51) gives τ 7·90. The predicted value for methyl attached to an amino group, as CH_3NH–, is τ 7·85, but this function is precluded by previous reasoning.

The fragments C_6H_5·, ·CONH·, ·CH_2CH_2·, and CH_3CO·, which together constitute the molecule of the unknown compound, can be fitted together in only one way if near-equivalence of the two methylene groups is to be achieved. This is as laevulinic anilide shown on p. 39.

---

* Note the offset.

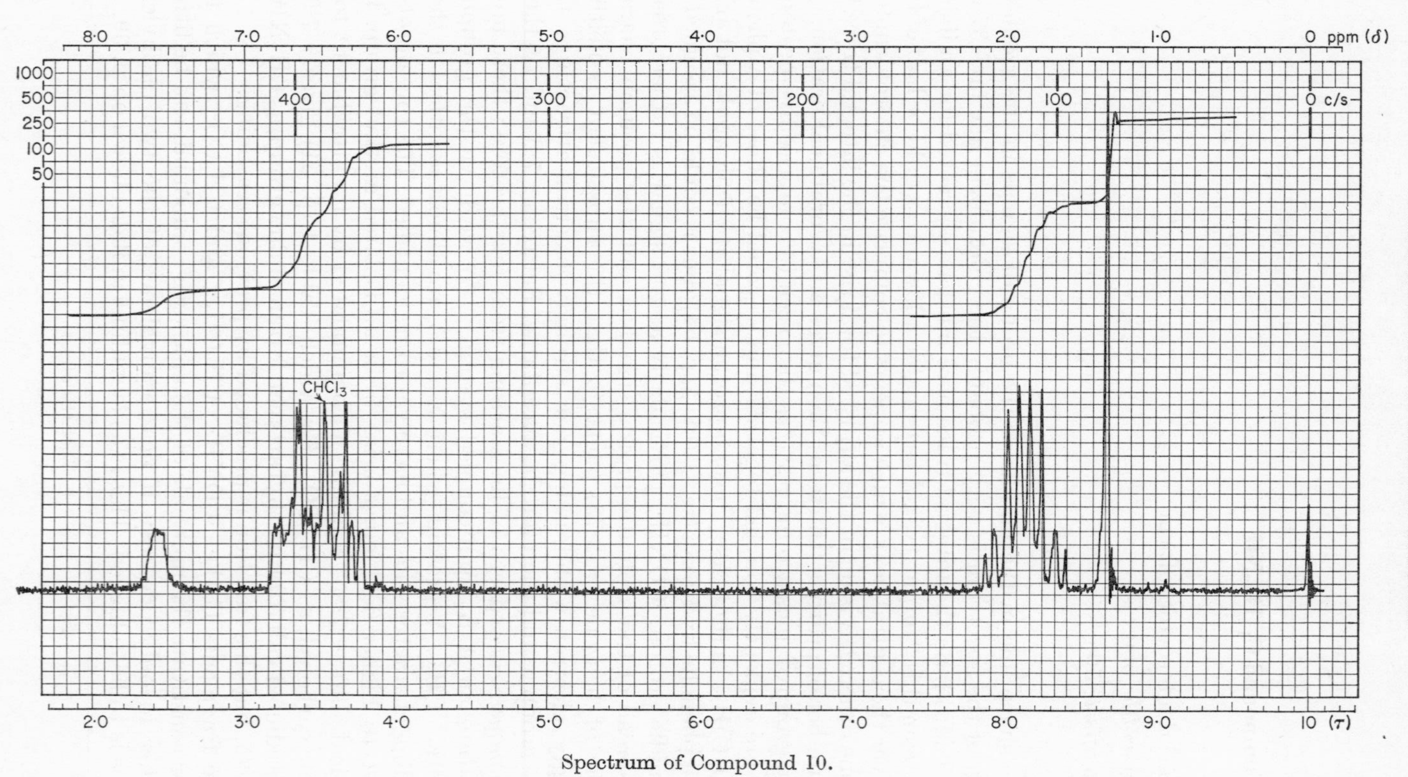

Spectrum of Compound 10.

Summary

Signal (c/s)	Position (ppm) δ	τ	Intensity	Multiplicity	Assignment
129	2·15	7·85	3	Singlet	$CH_3CO\cdot$
144 to 180	~2·7	~7·3	4	Symmetrical pattern (A_2B_2)	X—CH_2CH_2—Y (where X and Y are different carbonyl functions)
410 to 460	~7·3	~2·7	5	Complex	five protons of a phenyl group, $C_6H_5\cdot$
505	8·42	1·58	1	Broad	$\cdot NHCO\cdot$

This substance is

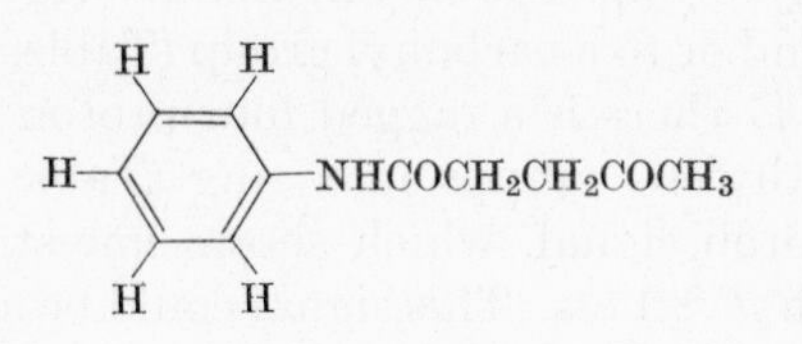

11. Compound $C_{13}H_{20}O$

Solvent: carbon tetrachloride
Sweep width: 500 c/s
Sweep offset: zero

In the olefinic region of the spectrum, at τ 4·02 and 2·87*, there are one-proton doublets which together form an AB pattern. The coupling constant is 16 c/s, so that the two protons must be *trans* about a double bond (Table 6, p. 57). The broadening of the lines of the lower-field doublet signifies *weak* coupling of the proton to others, which therefore must be several bonds away. The low-field positions of the signals suggest (Table 4, p. 54) that the protons are situated on a conjugated system.

At τ 8·93 there is a very sharp six-proton singlet, indicative of an isolated *gem*-dimethyl group. The chemical shift suggests that the methyls are β to a double bond or to a carbonyl group (Tables 1 and 2, pp. 51 and 52). Centred at τ 8·47 there is a ragged four-proton signal, which could arise from two methylene groups in a ring (Table 1). Then at τ 8·25 there is a three-proton signal, which shows fine-structure. It appears to be a quartet, with $J \approx 1$ c/s. This signal could be assigned to a methyl group attached to a fully substituted double bond (Table 1), the attached groups bearing three hydrogen atoms between them, to account for the small quadruplet splitting (see Table 6, p. 57). (Protons attached directly to this double bond would produce much larger splitting of the methyl signal.)

So far, then, there are good indications that the molecule contains the fragments:

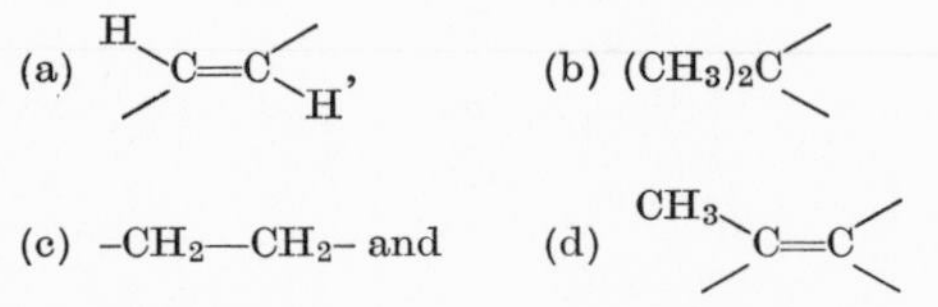

with (a) and (d) joined together and (b) attached, but with (c) separate. The residue, yet to be characterized is C_3H_5O.

Immediately to low-field of the foregoing signal at τ 8·25, there are further signals which, as the integral trace shows, arise from the remaining five protons. The sharp singlet at τ 7·82 appears to have an intensity between three and four. Careful inspection suggests that it is really a three-proton signal (necessarily, then, from a methyl group) superimposed on the lower-field one-third of the adjacent signal, this being a broadened two-proton triplet. This last signal, centred at τ 7·95,

---

* Note that the chart paper used for this spectrum shows only a δ scale.

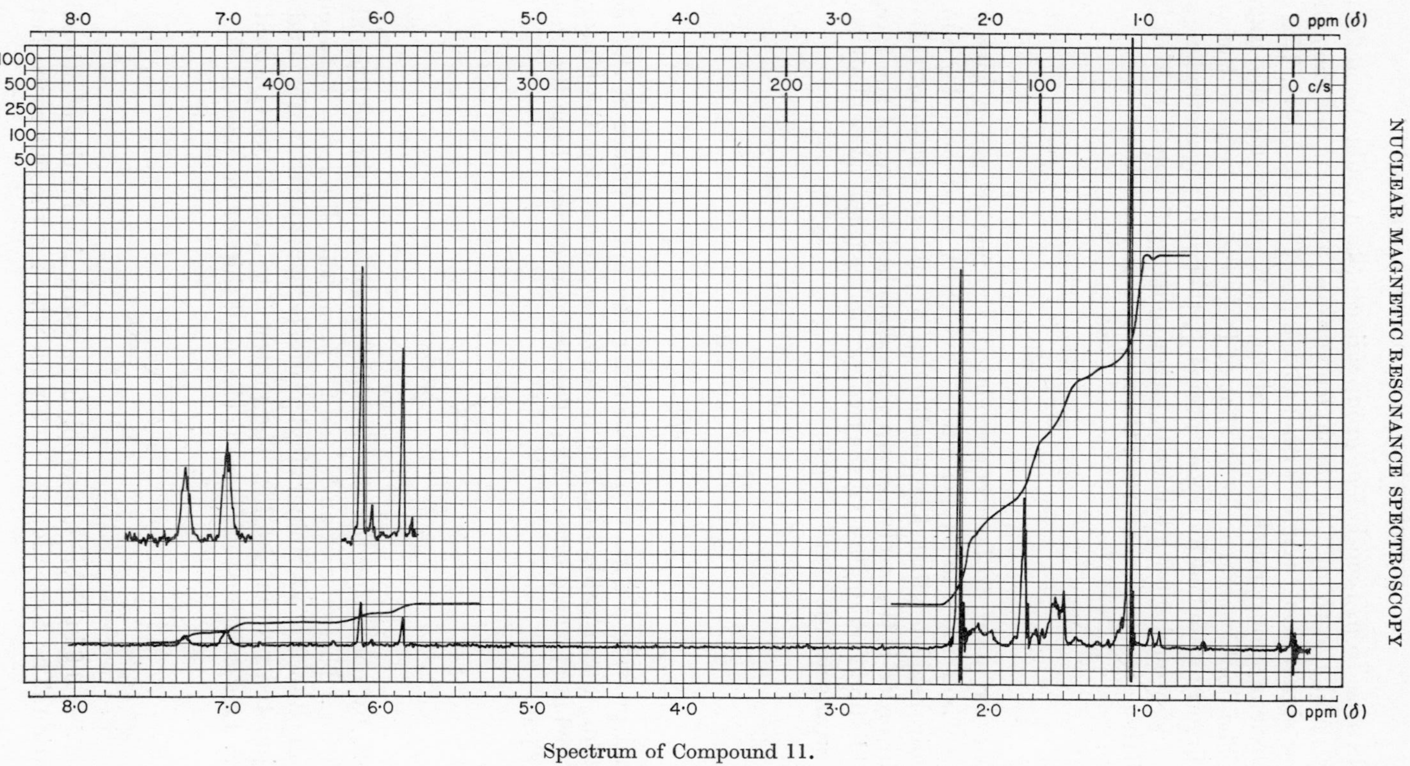

Spectrum of Compound 11.

could arise from a methylene group adjacent to a double bond (Table 1, p. 51) and also to another methylene group to account for the splitting of the signal into a triplet with $J \approx 5$ c/s (see Table 6, p. 57). This other methylene group must be one of the two comprising the fragment (c). The double bond must be that of fragment (d) because this carries no directly attached hydrogen atoms. [Otherwise, the methylene signal at τ 7·95 would show additional, larger splitting (Table 6, r. 57).] Hence a further structural feature of the molecule is:

H_2C — (c), (d)

There remain, to be accommodated, the methyl group with τ 7·82, one carbon and one oxygen atom—evidently as an acetyl group. The chemical shift is a little low—Table 1 (p. 51) gives τ 7·90—but this would be in order if the acetyl group were attached to one of the foregoing double-bond fragments. Of the two, (a) above is the more likely because the arrangement

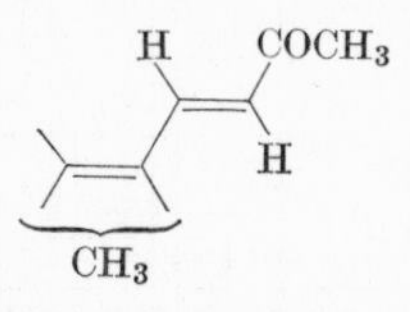

would provide a good explanation for the low τ values of the olefinic protons. From Table 4 (p. 54), it can be estimated that the proton α to the carbonyl would have τ 3·95. This value is derived from the chemical shift for

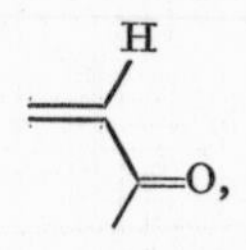

τ 4·2, by correcting for the deshielding effect of the additional conjugation. This correction is $-0{\cdot}25$ ppm, the difference between the chemical shift of a terminal vinyl proton (τ 5·35) and that of an analogous conjugated proton (τ 5·11). The proton β to the carbonyl would have τ 2·9. This is estimated from the chemical shift of the proton

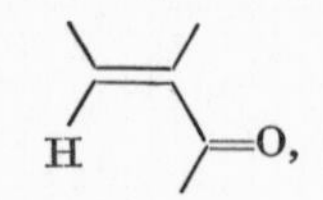

given as τ 3·8 in Table 4, by applying a correction for the additional conjugation. The correction is $-0{\cdot}9$ ppm which is the difference between the chemical shift of a non-terminal olefinic (τ proton 4·7) and that of an analogous conjugated proton (τ 3·8). These predicted values are very close to the observed ones.

To summarize, we have deduced that fragment (a) bears an acetyl group and is joined to (d), which in turn must be linked to (b) (the alternative attachment for (b) having been eliminated). Fragment (d) must also be linked through a methylene group to (c):

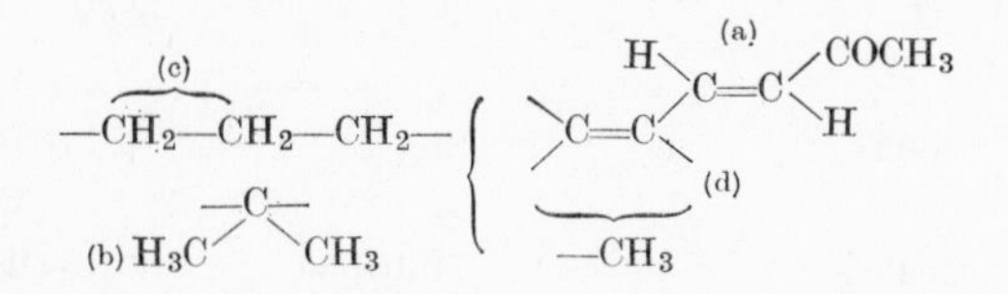

As fragment (c) cannot be adjacent to a double bond, (c) and (b) must be linked together. There are then four possible structures.

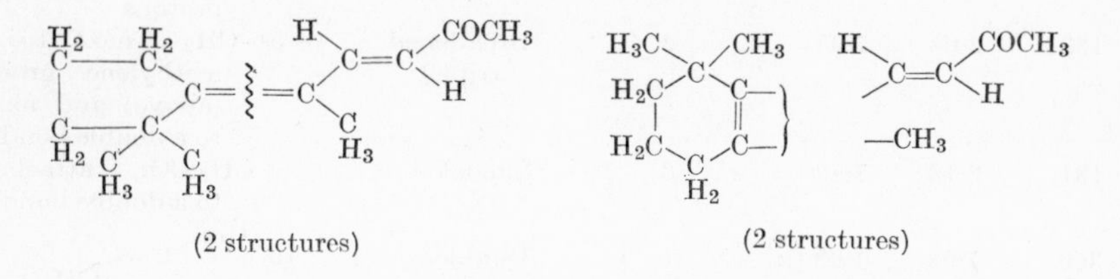

A firm decision between these could be obtained by chemical degradation, but can hardly be reached from the proton resonance data available here. It can be argued tentatively, however, that because the chemical shift of the *gem*-dimethyl group is lower than expected on the basis of it being β to a double bond, it must be close also to another double bond. This would suggest that of the above alternatives, the following is preferable:

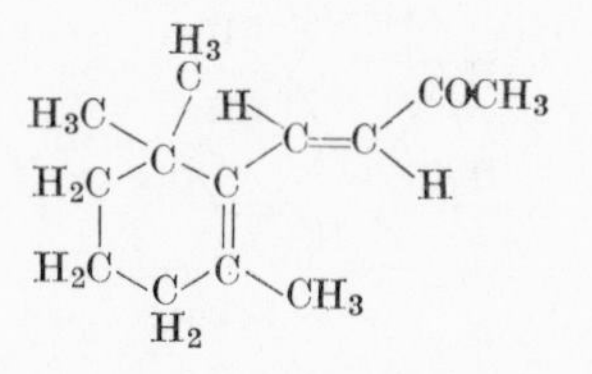

[A further tentative argument is that because in Example 2, p. 9, the chemical shift of the 4,5 methylene groups was about τ 8·32, a lower value than this might be expected for the remote methylene groups in a cyclopentylidene ring bearing an unsaturated substituent—there is less shielding in a five- than a six-membered ring (Table 1, p. 51). Therefore, as the observed shift is τ 8·47, a cyclohexenyl ring seems the more likely possibility.]

Summary

Signal (c/s)	Position (ppm) δ	τ	Intensity	Multiplicity	J (c/s)	Assignment
64	1·07	8·93	6	Singlet		$(CH_3)_2C<$, β to unsaturation
~92	1·53	8·47	4	Multiplet		$-CH_2-CH_2-$, in a ring.
105	1·75	8·25	3	Quartet	~1	$CH_3-C=C-$, coupled to three distant protons
123	2·05	7·95	2	Broadened triplet	~5	$-CH_2-$, next to a methylene group (above) and next to a double bond.
131	2·18	7·82	3	Singlet		$CH_3.CO.$, attached to a double bond
359	5·98	4·02 (B)	1	Doublet	16	$H_{(B)}C=CH_{(A)}$
428	7·13	2·87 (A)	1	Doublet lines broadened	16	with additional conjugation; $H_{(A)}$. coupled to distant protons

This substance is β-ionone, in the conformation[4]:

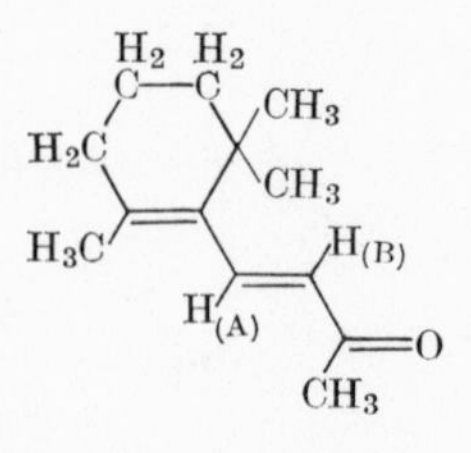

12. Compound $C_7H_6O_3S$

Solvent: deuterochloroform
Sweep width: 500 c/s
Sweep offset: 250 c/s

(*A*) Shows the effect on the broad resonance at ~720 c/s of shaking the solution with D_2O; (*B*) shows the effect on the broad resonance at 280 c/s of adding 1 drop of conc. hydrochloric acid.

The tetramethylsilane line is shown at τ 10, but the rest of the spectrum is displaced 250 c/s to lower-field. So 250 c/s has to be added to the apparent line-positions in order to obtain the correct values. At 437·5 c/s from T.M.S. there is a very sharp singlet. This is the line from chloroform present in the deuterochloroform.

Around the chloroform line, in the region of the spectrum characteristic of olefinic and aromatic protons, there are four one-proton signals. One of the protons responsible is isolated because the signal is a single sharp line—that at τ 2·22. The other three are coupled together as two pairs: this follows from the splitting pattern. Two of the signals, those at τ 3·05 and 2·37, are doublets with splittings of 3·5 and 2·0 c/s, respectively. Each of the protons responsible must be coupled to one other proton, which in each case must give rise to a signal at higher field, judging from the relative intensities of the lines of the doublet signals. To higher-field, there are not two doublet signals, but only a single one-proton signal at τ 3·40 which, however, shows *both* of the splittings previously encountered. It is a double-doublet. The proton responsible must therefore be coupled to each of the other two, and so we deduce that the molecule of the dissolved compound contains the following atomic arrangement:

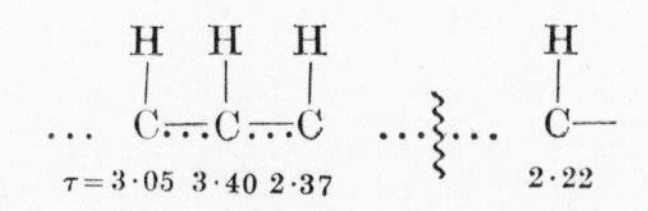

where the carbon atoms are part of an aromatic or olefinic system. (The signal at τ 3·40 appears as *four* lines because the proton is coupled *unequally* to each of *two* other protons: the splitting rule $N = (n_1 + 1)(n_2 + 1)$ etc.* therefore applies, where $n_1 = n_2 = 1$. We can imagine that

* This expression is equivalent to application of the simple "splitting rule" in stages (see p. 3). It gives the maximum number of lines which could arise, but it does not take into account possible overlap of lines which can occur with certain relative values of the different coupling constants.

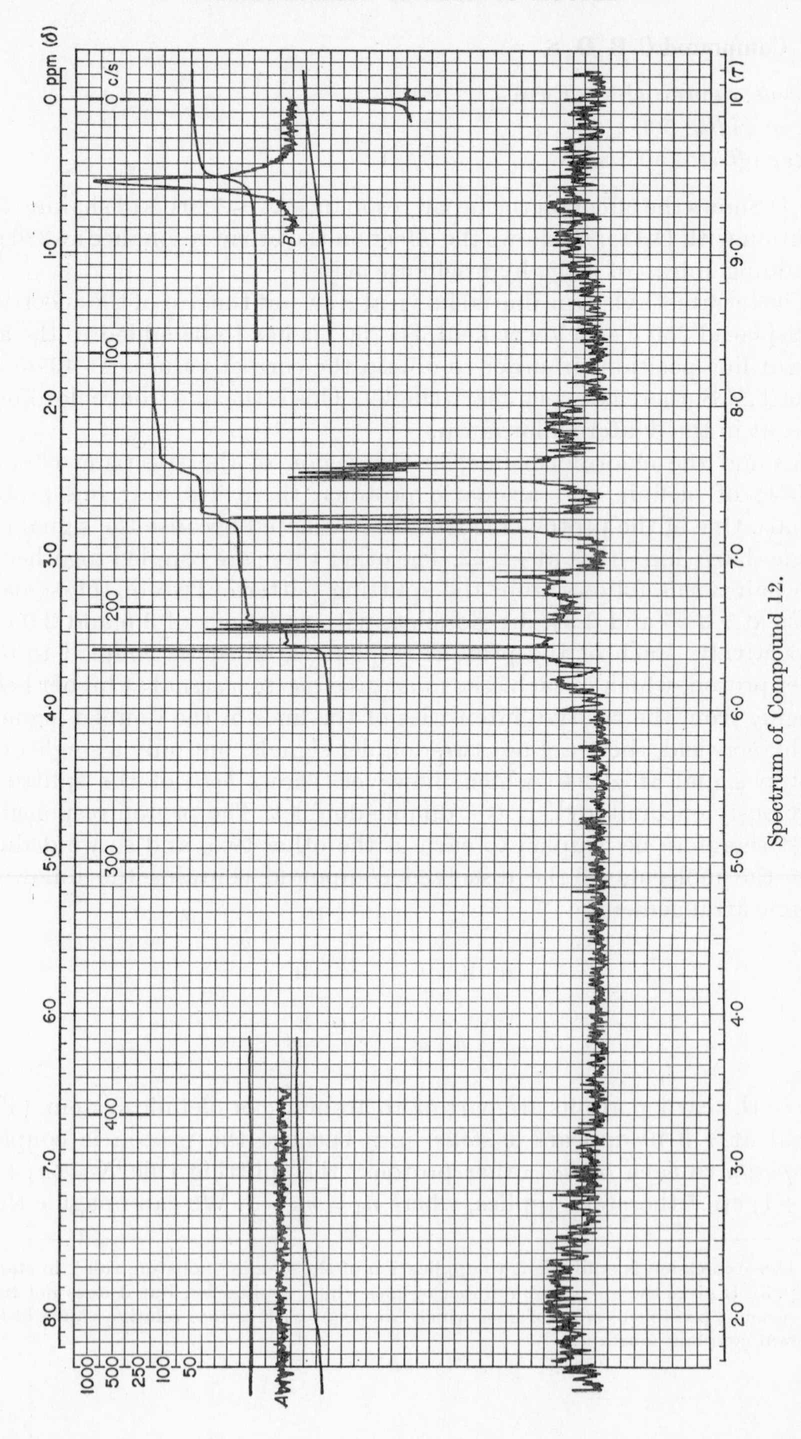

Spectrum of Compound 12.

the origin signal at τ 3·40 is split into two lines by coupling with one proton, and then each of the lines is split into two by the coupling with the other proton, so that four lines (which do not overlap) are produced. The signal is described as a double-doublet, not a quartet, because the lines are not equally spaced, which is a direct result of there being two different coupling constants.)

Near 720 c/s from T.M.S., i.e. at $\tau - 2{\cdot}0$, there is a broad one-proton signal. Its position, and the fact that it disappears immediately when the solution is shaken with heavy water, suggest that it could arise from the proton of a carboxyl group (Table 5, p. 56) or possibly an enol function. Strongly acidic functions give sharp proton-resonance lines, so the possibility of a sulphonic acid grouping, $-SO_3H$, being present can be dismissed.

There is another broad one-proton signal near τ 5·34 which sharpens up (and moves a little) when a *trace* of hydrogen chloride is added to catalyse exchange. The values in Table 5 (p. 56) suggest that this signal might arise from a phenolic hydroxyl, but then the line should be reasonably sharp. A satisfactory assignment would be to a thiol group, in the form of an enol function,

$$\gtrdot C\text{—SH}$$

this gives a signal to lower-field of the region normally characteristic of a simple thiophenol.

It appears, then, that the molecule may comprise the following fragments

$$\overset{\text{H}}{\underset{|}{\text{C}}}\cdots\overset{\text{H}}{\underset{|}{\text{C}}}\cdots\overset{\text{H}}{\underset{|}{\text{C}}},\quad \cdots\overset{\text{H}}{\underset{|}{\text{C}}}\text{—},\quad -CO_2H,\quad \rangle\text{—SH}$$

together with C and O.

The magnitudes of the couplings between the CH groups suggest that these groups cannot be linked by normal olefinic double bonds (Table 6, p. 57). Moreover, the coupled protons cannot be *meta* to one another on a benzene ring (cf. Table 6), because then there would be only one coupling constant and the signal from each proton would appear as a triplet. However, from the three CH groups and the oxygen and carbon atom, a furan ring can be constructed. This would accommodate both the couplings and the chemical shifts. The lowest-field proton would be adjacent to the ring-oxygen atom, as follows.

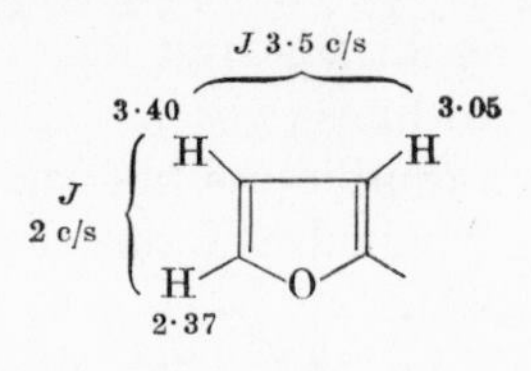

The lowering of the chemical shifts from the values for the protons in furan itself (Table 4, p. 54) could result from an unsaturated (electron-withdrawing) substituent. This would have to be constructed from the remaining molecular fragments:

H
|
C– , $-CO_2H$ and >–SH.

Two possible structures then emerge, that of α-thiol-β-2-furylacrylic acid (I) and its positional isomer β-thiol-β-2-furylacrylic acid (II)

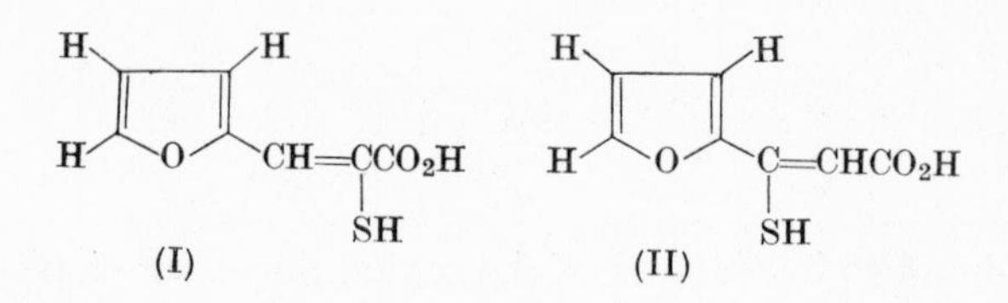

A decision between these can be reached by estimating the chemical shift of the olefinic proton in the side chain. For structure (II) we start with τ 5·1 as the chemical shift for the terminal proton of a conjugated system (Table 4, p. 54). The effect of the α-carbonyl may be minus 1·1 ppm (τ 8·50 $-$ τ 7·43, from Table 1, p. 51), and that of the β-thiol and the β-furyl ring may be minus 0·5 ppm (comprising 0·2 and 0·3 ppm, respectively, estimated from Table 2, p. 52). The chemical shift of the side-chain olefinic-proton in (II) might then be near

$$\tau\ 5{\cdot}1 - 1{\cdot}1 - 0{\cdot}5 = \tau\ 3{\cdot}5.$$

For structure (I), we start with τ 3·8 as the chemical shift of a non-terminal proton in a conjugated system and adjust this for the presence of an aromatic-type ring α to it: the adjustment may be minus 1·4 ppm (τ 8·50 $-$ 7·13, from Table 1, p. 51). A further adjustment for the presence of a carboxyl and a thiol group β to the proton, would be minus 0·3 ppm (estimated as 0·1 and 0·2 ppm, respectively, from Table 2, p. 52). Thus, the chemical shift of the side-chain olefinic-proton in (I) might be near τ 3·8 − 1·4 − 0·3 = τ 2·1. This is much closer to the observed value, τ 2·22, than the previous estimate. Hence it may be concluded tentatively that structure (I) is correct. Rough calculations of chemical shifts along the above lines are open to criticism, but are frequently helpful. If such an estimated chemical shift leads to a conflict with other structural evidence, or if other evidence is lacking, then it is essential to determine the proton chemical shifts for closely related model compounds, before reaching structural conclusions.

Summary

Signal (c/s)	Position (ppm) δ	τ	Intensity	Multiplicity	J (c/s)	Assignment
~280	4·66	5·34	1	Broad, sharpened by H^+		$>$CSH
396	6·60	3·40	1	Double-doublet	3·5 2·0	$H_{(C)}$ } of $H_{(A)}$, $H_{(C)}$, $H_{(B)}$ on =C—C=C= where $J_{AB} = 0$
417	6·95	3·05	1	Doublet	3·5	$H_{(B)}$
457·5	7·63	2·37	1	Doublet	2·0	$H_{(A)}$
467	7·78	2·22	1	Singlet		H—C$<$
~720	12·0	−2·0	1	Broad		$-CO_2H$

The structure is

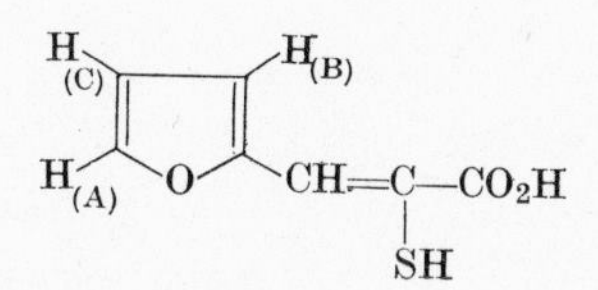

Proton Magnetic Resonance Correlation Tables

The following τ values are mostly average values from several compounds and sources. Discrepancies from values observed in particular cases might be as much as $\pm 0{\cdot}5$ ppm.

TABLE 1

τ Values (approx.) for the protons of CH_3, CH_2 and CH groups (attached to various groups X, and to saturated hydrocarbon residues, R, R′, etc.)

X	CH_3X	$R'CH_2X$*	$R'R''CHX$
–R	9·10	8·75	8·50
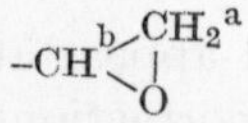 $-CH^{b}(CH_2^{a})O$ (epoxide)	8·68	a 7·6, 7·3	b 7·0
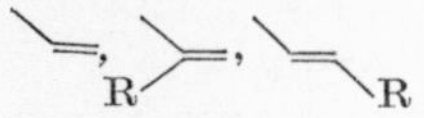 (alkenyl: ⟍═, R⟋⟍═, ⟍═⟍R)	8·31	8·05	7·4
 ⟍═—═—═ etc. (i.e. end-of-chain)	8·18		
 ═⟍═—═ etc. (i.e. in-chain)	8·03	7·8	
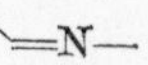 ⟍═N—	8·0		
–COOR	8·00	7·90	
–CN	8·00	7·52	
$-CONH_2$, $-CONR_2$	7·98	7·95	
–COOH	7·93	7·66	7·43
–COR	7·90	7·60	7·52
–SH, –SR	7·90	7·60	
$-NH_2$, $-NR_2$	7·85	7·50	7·13
–I	7·84	6·85	5·80
–CHO	7·83	7·8	7·6
–Ph‡	7·66	7·38	7·13
–Br	7·35	6·66	5·90
–NHCOR, –NRCOR‴	7·1	6·7	6·5
—Cl	6·98	6·56	5·98
–OR	6·70	6·64	6·20
$[-\overset{+}{N}R_3]$	6·67	6·60	

[continued]

TABLE 1—*cont.*

X	CH_3X	$R'CH_2X$*	$R'R''CHX$
–OH	6·62	6·44	6·15
–OCOR	6·35	5·85	4·99
–OPh	6·27	6·10	6·0
–OCOPh	6·10	5·77	4·88
–F	5·74	5·65	
$–NO_2$	5·67	5·60	5·40

* If present in 3-membered ring, +1·05
If present in 5-membered ring, −0·25 }
If present in 6-membered ring, −0·20 } except for a CH_2 next to >C=O, when the
If present in 7-membered ring, −0·30 } sign is changed.

‡ *o*-, *m*-, *p*-substituents, { electron-withdrawing, 0 to −0·2; electron-donating, 0 to +0·1 }

TABLE 2

(a) CH_3, CH_2, CH positioned β to the X groups of Table 1 appear at values 0·05 to 1·0 *lower* than shown above. The additional corrections are as follows.

X	β-Shifts CH_3—C—X	CH_2—C—X	CH—C—X
–C=C	−0·1	−0·05	
–COOH, –COOR	−0·25		
–CN		−0·4	
$–CONH_2$	−0·23		
–COR, –CHO	−0·2		
–SH, –SR	−0·45	−0·3	
$–NH_2$, $–NR_2$	−0·1	−0·05	
–I	−1·0	−0·5	−0·4
–Ph	−0·35	−0·3	
–Br	−0·8	−0·6	−0·25
–NHCOR	−0·1		
–Cl	−0·6	−0·4	−0·02
–OH, –OR	−0·27	−0·1	
–OCOR	−0·37		
–OPh		−0·35	
–F		−0·2	
$–NO_2$		−0·8	
–C—Y†	0·0 to −0·15		

† Y is any common function (–OH, –CO, –N<, Hal., etc.)

TABLE 2—*cont.*

(b) CH_3 (etc.) attached to naphthyl and higher hydrocarbons, and to α-pyridyl, pyrimidyl, furyl, thienyl, thiazolyl, etc., appears at values 0·05 to 0·65 *lower* than in toluene (7·66).

CH_3 (etc.) attached to pyrryl, β- or γ-pyridyl comes at values 0·05 to 0·3 *higher*.

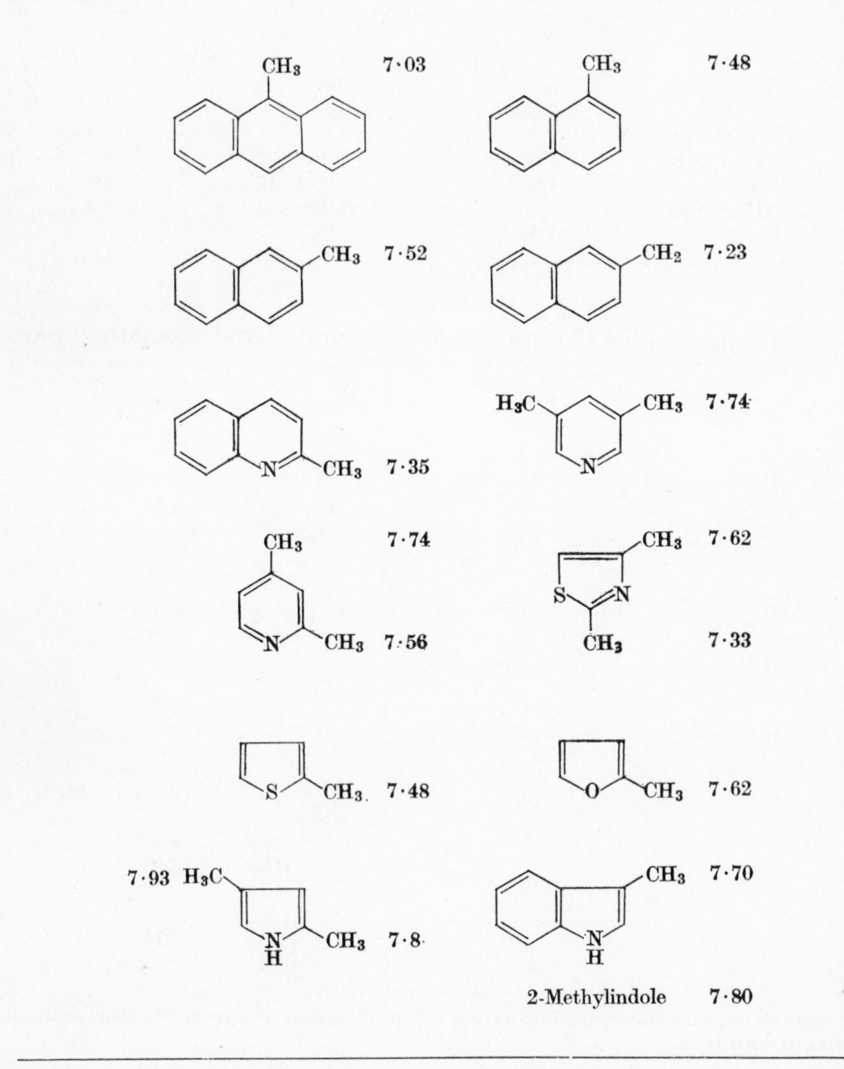

TABLE 3

τ Values (approx.) for the protons of CH_2 (and CH) groups bearing more than one functional substituent.

(Modified Shoolery Rules)

For $H_2C\begin{matrix}X^1\\X^2\end{matrix}$ $\tau_{CH_2} = 8{\cdot}75 - \Sigma a.$

Less accurate for $HC\begin{matrix}X^1\\X^2\\X^3\end{matrix}$

X	*a*	X	*a*
—=	0·75	–Ph	1·3
—≡	0·90	–Br	1·9
–COOH, –CO·OR	0·7	–Cl	2·0
–CN, –CO·R	1·2	–OR, –OH	1·7
–SR	1·0	–O·CO·R	2·7
$–NH_2$, $–NR_2$	1·0	–OPh	2·3
–I	1·4		

TABLE 4

τ Values (approx.) for H attached to unsaturated and aromatic groups

H—≡—	8·20*	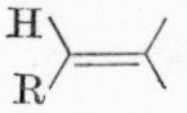 H(R)C=C<	4·7
H—≡—C(—)—OH	7·6*	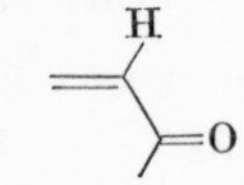 CH_2=C(H)—C(=O)—	4·2
H—≡—≡— etc.	7·3*		
H—≡—Ph	7·07*	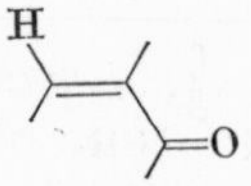 H(—)C=C(—)—C(=O)—	4·0
H—≡—CO—	6·83*		
H_2C=C(R)(R′)	5·35	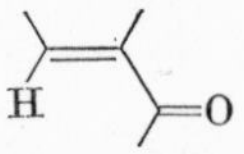 (—)(H)C=C(—)—C(=O)—	3·8
H_2C=—=— etc.	5·1	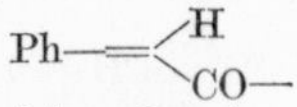 Ph—CH=CH—CO— (*cis* or *trans*)	3·4
H(R)C=—OR′ (acyclic)	5·0	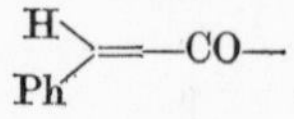 H(Ph)C=—CO—	2·2
		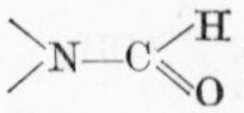 >N—C(H)=O	2·15

* Signals moved to lower-field by a trace of pyridine; they disappear when the solution is shaken with D_2O.

TABLE 4—*cont.*

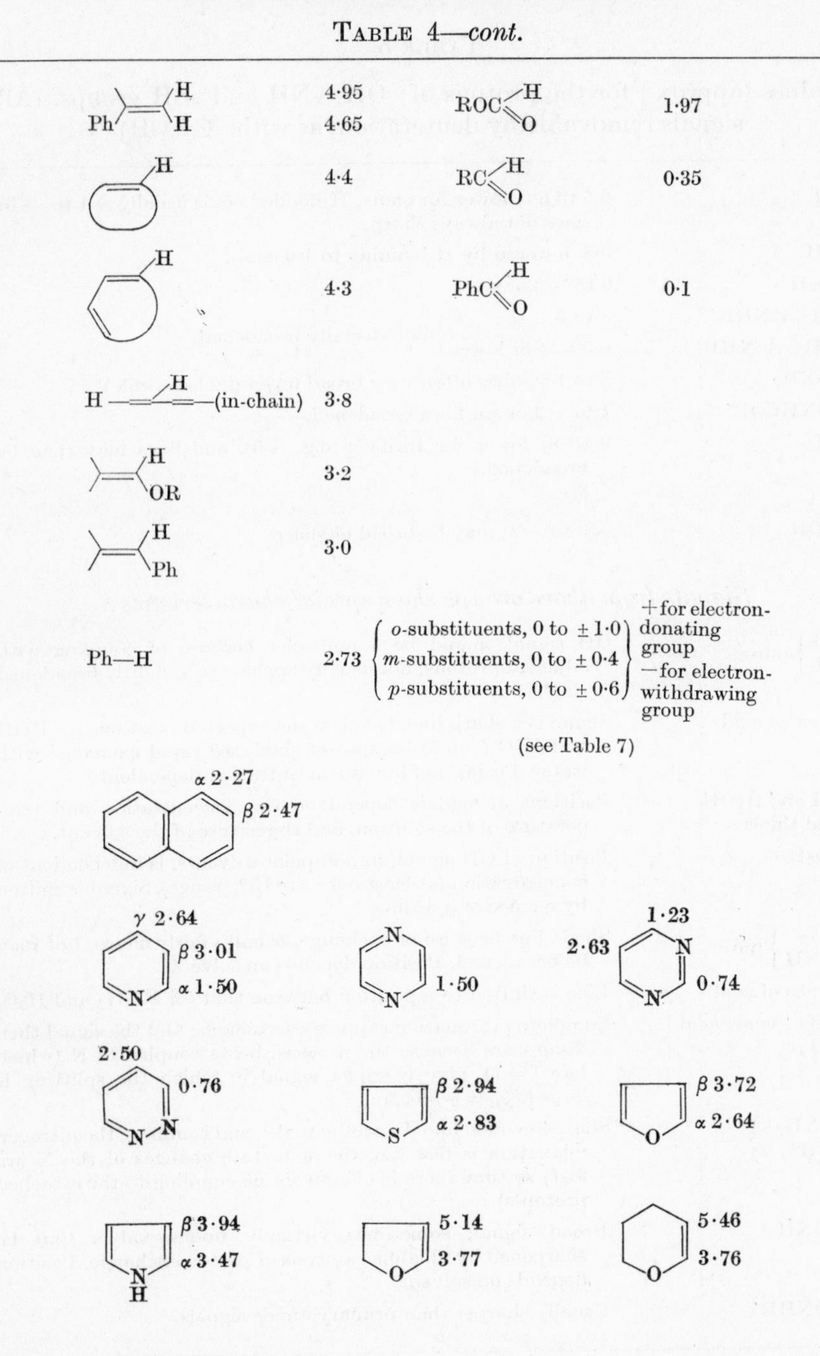
H
Ph H
4·95
4·65
ROC H
O
1·97
H
4·4
RC H
O
0·35
H
4·3
PhC H
O
0·1
H
H (in-chain) 3·8
H
OR
3·2
H
Ph
3·0
Ph—H
2·73
o-substituents, 0 to ± 1·0
m-substituents, 0 to ± 0·4
p-substituents, 0 to ± 0·6
+for electron-donating group
−for electron-withdrawing group
(see Table 7)
α 2·27
β 2·47
γ 2·64
β 3·01
α 1·50
N
N
N
1·50
1·23
2·63
N
N
0·74
2·50
0·76
N
N
β 2·94
α 2·83
S
β 3·72
α 2·64
O
β 3·94
α 3·47
N
H
5·14
3·77
O
5·46
3·76
O

TABLE 5

τ Values (approx.) for the protons of –OH, –NH and –SH groups. (All signals removable by deuteration, as with –C≡CH)

ROH	9·5 to 6·0, lower for enols; H-bonded enols usually −1 to −6: lines not always sharp.
$\bar{A}$rOH	5·5, lowered by H-bonding to 1·0 or so.
RCO_2H	0 to −3 or so.
RNH_2, RNHR′	8 to 5 } lines usually broadened.
$\bar{A}rNH_2$, $\bar{A}$rNHR′	6·5 to 4 or lower } lines usually broadened.
$RCONH_2$	5 to 1·5; lines often very broad (even unobservable).
RCONHCOR′	1 to −2 or so; lines broadened.
RSH	9 to 8; lower for enthiols (e.g. ~5), and lines may then be broadened.
$\bar{A}$rSH	~6·5.
=NOH	~0 to −2; may be broad or sharp.

Signals from above groups show special characteristics:

ROH, RSH } pure	OH signal should be a multiplet because of coupling with adjacent protons, but usually appears as a slightly broadened singlet.
+ trace of acid	Signal is a sharp line, between the expected positions for ROH and H_2O (~5·3) because of catalysed rapid exchange with water. Position of line is concentration-dependent.
ROH and $\bar{A}$rOH and thiols	Positions of signals depend on the concentration and temperature of the solution, and the nature of the solvent.
RCO_2H	Position of OH signal, in non-polar solvents, is independent of concentration (at least over 5 to 10% range). Signal is shifted by a trace of pyridine.
RNH_2, RR′NH } pure	Single line because of exchange, usually fairly sharp, but may be broadened. Position depends on solvent.
+ trace of acid	Line is shifted to a position between that for RNH_2 and H_2O.
RNH_2 + conc. acid, RNH_3^+	Complete protonation suppresses exchange and the signal then disappears because the protons, being coupled to N (which has $I = 1$), give a triplet signal in which the splitting is large ($J_{NH} = \sim 50$ c/s).
$RR'NH_2^+$, R_3NH^+	Sharp low-field line. In protonated s- and t-amines, the nitrogen relaxation is fast (i.e. the spin-state changes of the N are fast) so that there is effectively no coupling to the attached proton(s).
$RCONH_2$	Broad signal, sometimes virtually unobservable. Can be sharpened by alkaline catalysis of proton exchange. Position depends on solvent.
RCONHR′	Usually sharper than primary amide signals.

TABLE 6

Proton-proton spin coupling constants

Function	J_{ab} (c/s)
$>C<^{H_a}_{H_b}$ (*gem*)	-10 to -18 depending on the electronegativities of the attached groups (Ref. 2d, e and f).
$>CH_a—CH_b<$ (*vic*)	0 to 12, depending on dihedral angle ϕ (8 at 0°; 0 at 85°; 9·2 (Ref. 2g) to 11 (Ref. 1d) at 180° Ref. 2g gives: For $0° \leqslant \phi \leqslant 90°$, $J = 8{\cdot}5\ (\cos^2\phi) - 0{\cdot}28$; for $90° \leqslant \phi \leqslant 180°$, $J = 9{\cdot}5(\cos^2\phi) - 0{\cdot}28$.
$>C{=}C<^{H_a}_{H_b}$	-3 to $+7$ (Ref. 2d, e, f).
$H_a>C{=}C<H_b$ (*cis*)	5 to 14
$H_a—C{=}C—H_b$ (*trans*)	11 to 19
$>C{=}C(H_b)—C—H_a$	4 to 10
$H_a—C{=}C—C—H_b$ (*cis* or *trans*)	0 to -2 (0 to $+1$ when the "double bond" is part of an aromatic ring).
$>C{=}CH_a—CH_b{=}C<$	10 to 13
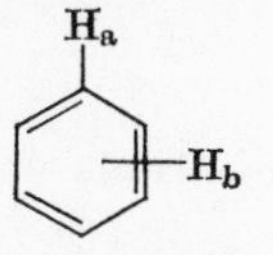 Benzene ring, H_a, H_b	*ortho*, 7 to 10 *meta*, 2 to 3 *para*, 1 or less, often unobservable.
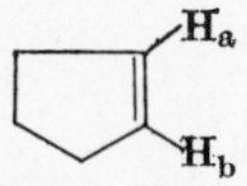 Cyclopentene ring, H_a, H_b on double bond	3 to 6
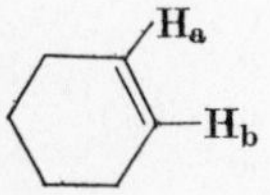 Cyclohexene ring, H_a, H_b on double bond	9 to 14
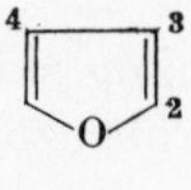 Furan (positions 2, 3, 4, O)	J_{23} 1·8, J_{34} 3·5, J_{24} 0·8, J_{25} 1·6
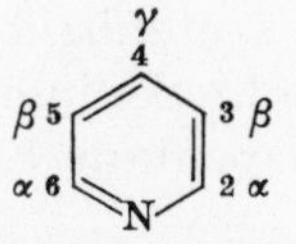 Pyridine (2 α, 3 β, 4 γ, 5 β, 6 α, N)	J_{23} 5·5, J_{34} 7·5 J_{24} 1·9, J_{35} 1·6, J_{26} 0·4 J_{25} 0·9

TABLE 7

Shifts (approx.) in the position of benzene protons (τ 2·73) caused by substituents

Substituent	*ortho*	*meta*	*para*
$-CH_3$	0·15	0·1	0·1
$-\!=$	−0·2	−0·2	−0·2
−COOH, −COOR	−0·8	−0·15	−0·2
−CN	−0·3	−0·3	−0·3
$-CONH_2$	−0·5	−0·2	−0·2
−CO·R	−0·6	−0·3	−0·3
−SR	−0·1	0·1	0·2
$-NH_2$	0·8	0·15	0·4
$-N(CH_3)_2$	0·5	0·2	0·5
−I	−0·3	0·2	0·1
−CHO	−0·7	−0·2	−0·4
−Br	0	0	0
−NHCOR	−0·4	0·2	0·3
−Cl	0	0	0
$-NH^+_3$	−0·4	−0·2	−0·2
−OR	0·2	0·2	0·2
−OH	0·4	0·4	0·4
−OCOR	−0·2	0·1	0·2
$-NO_2$	−1·0	−0·3	−0·4

TABLE 8

Approximate line positions of some protic solvents. The positions vary a little with solute

Solvent	τ	Solvent	τ
CF_3CO_2H	0·17	Dioxan	6·32
$HCONH_2$	2·15 (CH)	CH_3OH	6·53 (CH_3)
$CHCl_3$	2·73	$(CH_3)_2SO$	7·4
Benzene	2·73	$(CH_3)_2CO$	7·83
$CHBr_3$	3·15	CH_3CO_2H	7·90 (CH_3)
CH_2Cl_2	4·70	CH_3CN	8·00
H_2O	~5·3	Cyclohexane	8·57
CH_3NO_2	5·67	$(CH_3)_3COH$	8·73

The intense solvent-signal is always accompanied by spinning side-bands which occur in a region 0·5 to 1 ppm on each side. Nitrobenzene and pyridine obscure a region τ 3 to 1.

For solutions in H_2O and D_2O, dioxan (τ 6·32) and acetonitrile (τ 8·00) are good internal standards, better than t-butanol.[5] Sodium trimethylsilylpropanesulphonate (sodium 4,4-dimethyl-4-silapentane-1-sulphonate, DSS) is now widely used.[6]

Examples for Analysis

There now follow the spectra of six compounds for each of which is given the empirical formula with any relevant experimental data. It is suggested that the reader should evaluate the actual structures of the first five unknowns and assign the signals in the spectrum of the last example. The correct formulae for the first five will be found on p. 179.

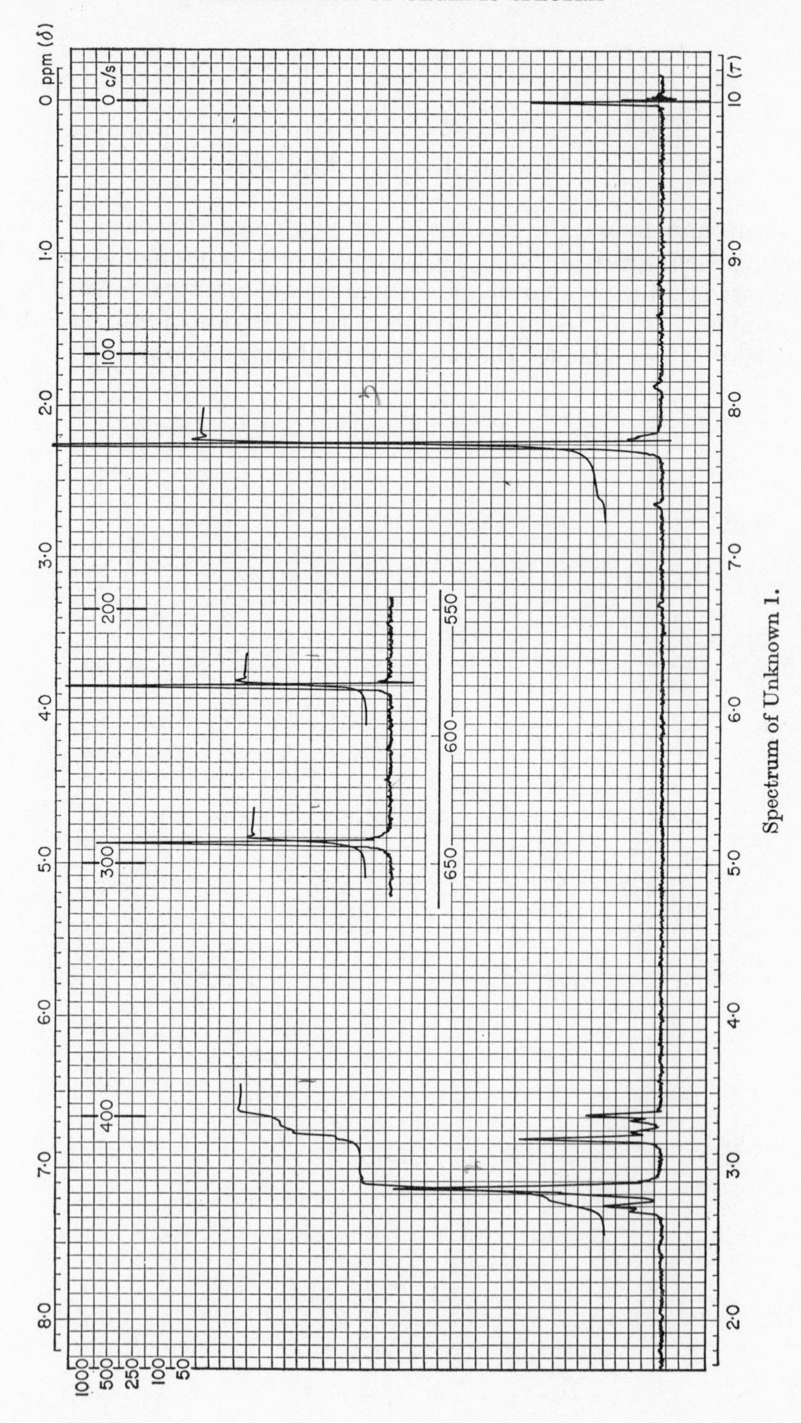

Spectrum of Unknown 1.

Unknown No. I

Compound $C_8H_8O_2$
Solvent: carbon tetrachloride
Sweep width: 500 c/s. The inset peaks are on the c/s scale

The lowest-field signal disappears on deuteration of the compound, by shaking the solution with slightly alkaline heavy water.

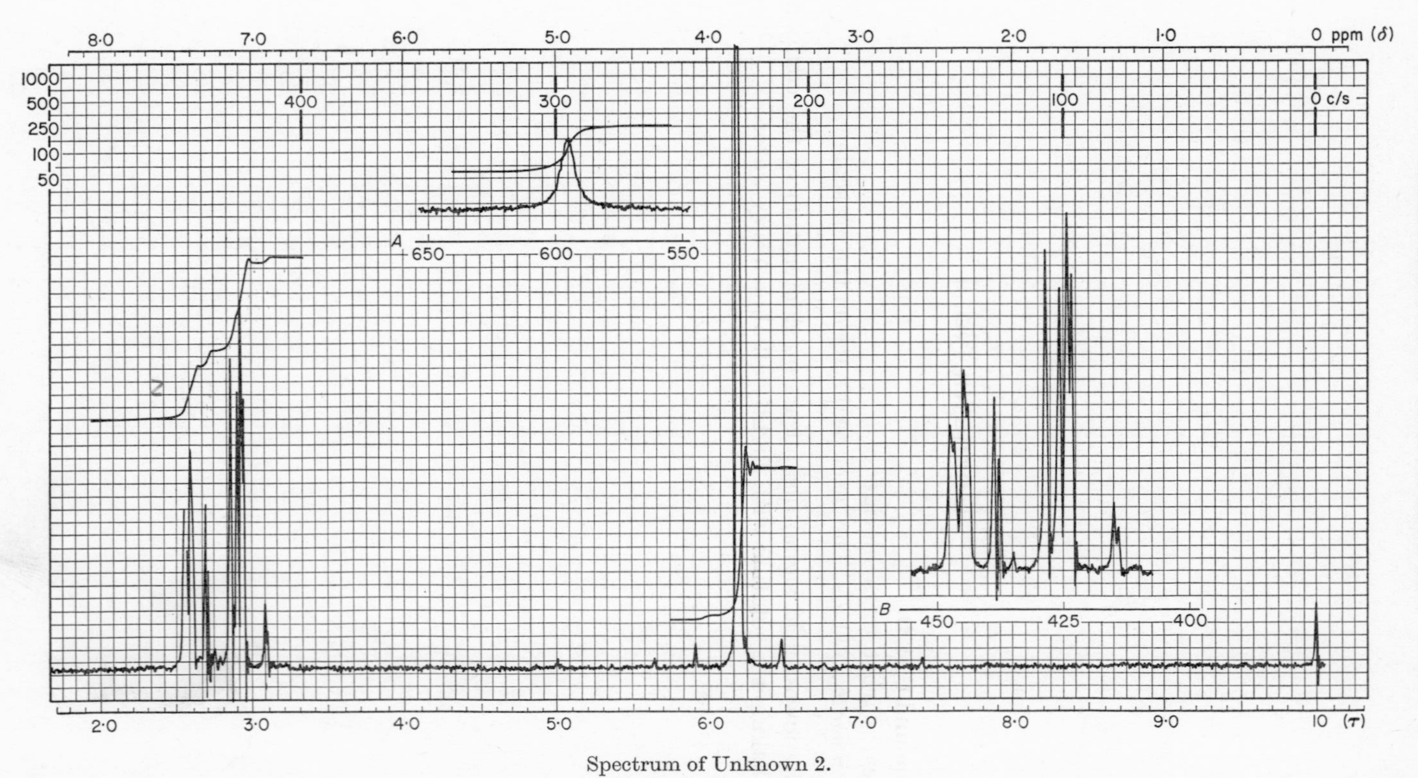

Spectrum of Unknown 2.

Unknown No. 2

Compound $C_7H_7O_4N$

Solvent: deuterochloroform
Sweep width: 500 c/s for main chart and inset *A*; 250 c/s for inset *B*
Sweep offset: zero for main chart
Insets *A* and *B* are shown on the c/s scale

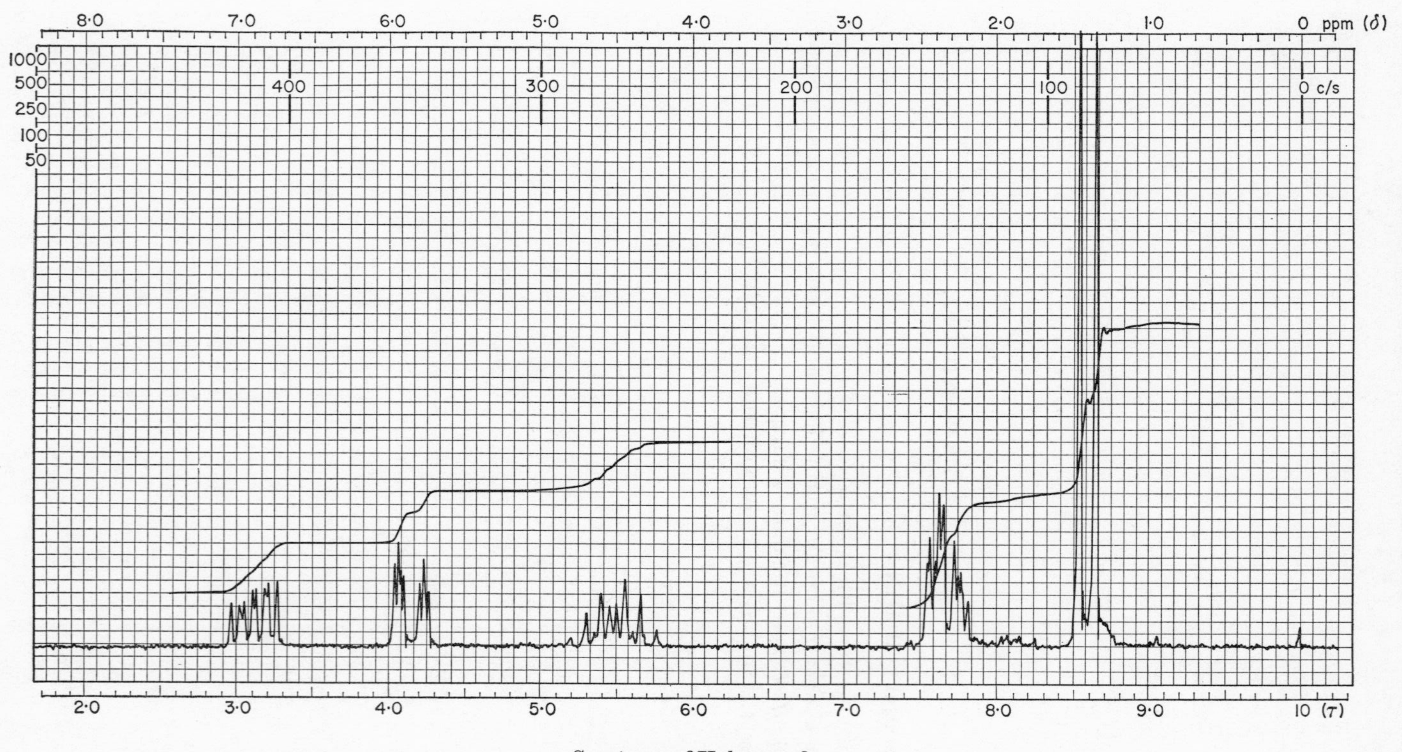

Spectrum of Unknown 3.

Unknown No. 3

Compound $C_6H_8O_2$

Solvent: carbon tetrachloride
Sweep width: 500 c/s
Sweep offset: zero
The inset is shown on the c/s scale

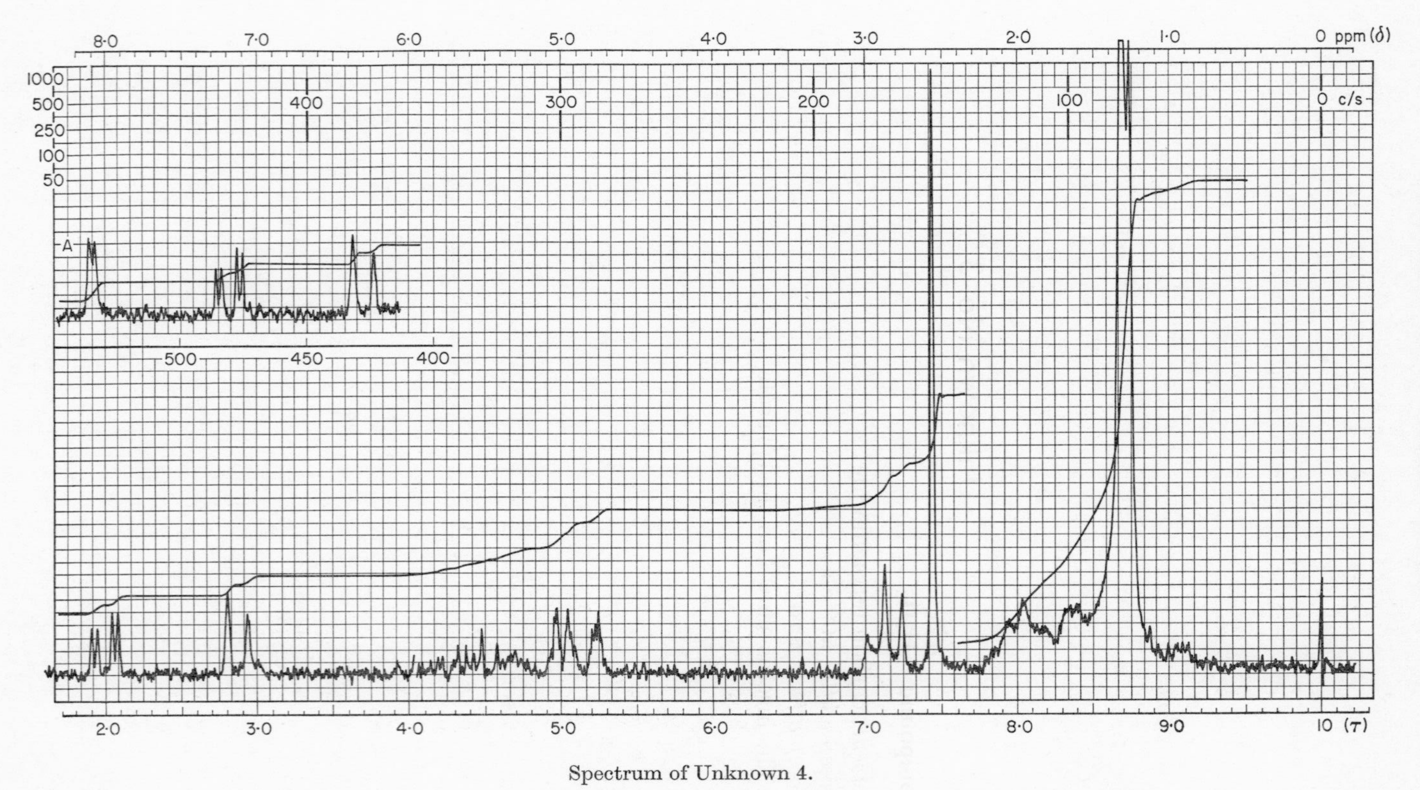

Spectrum of Unknown 4.

Unknown No. 4

Compound $C_{20}H_{31}ON$

Solvent: carbon tetrachloride
Sweep width: 500 c/s. Inset *A* is shown on the c/s scale
Sweep offset: zero for main chart.

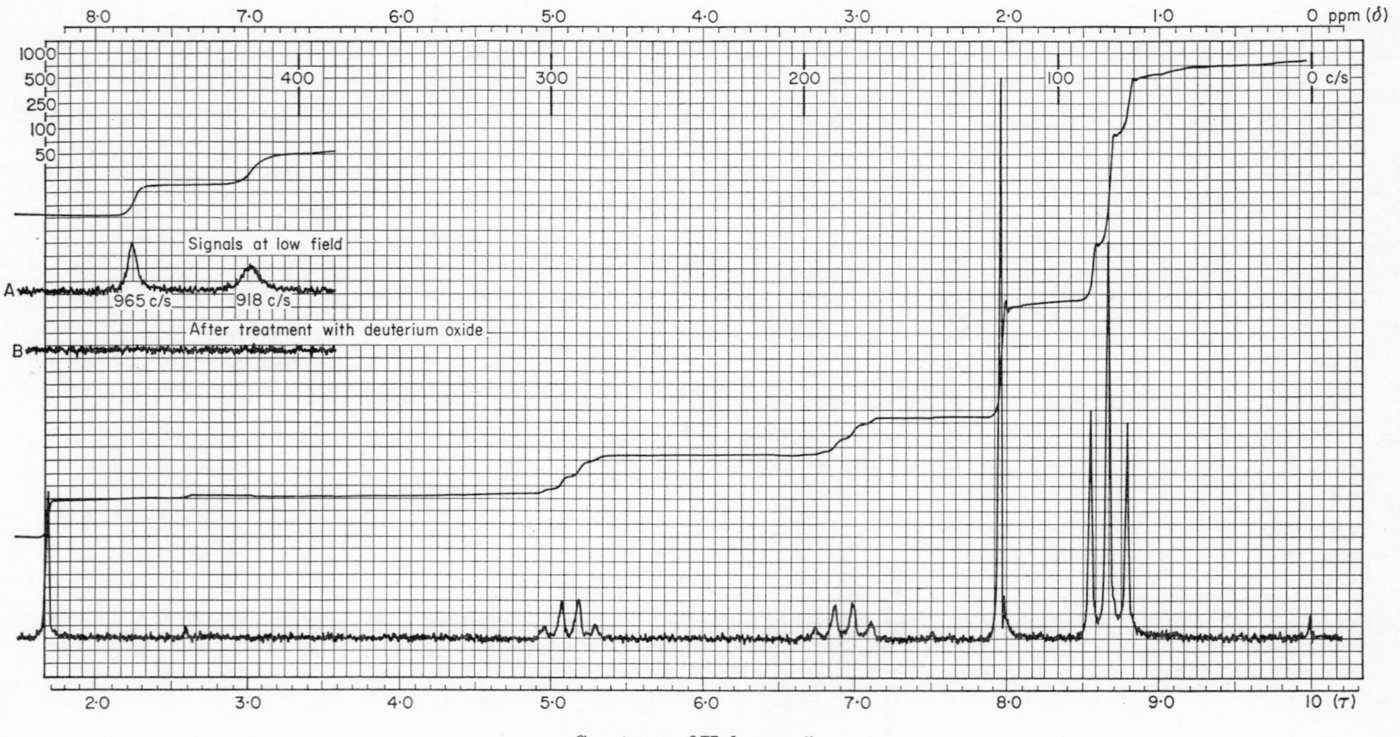

Spectrum of Unknown 5.

Unknown No. 5

Compound $C_{13}H_{14}O_5$

Solvent: deuterochloroform
Sweep width: 500 c/s
Sweep offset: zero

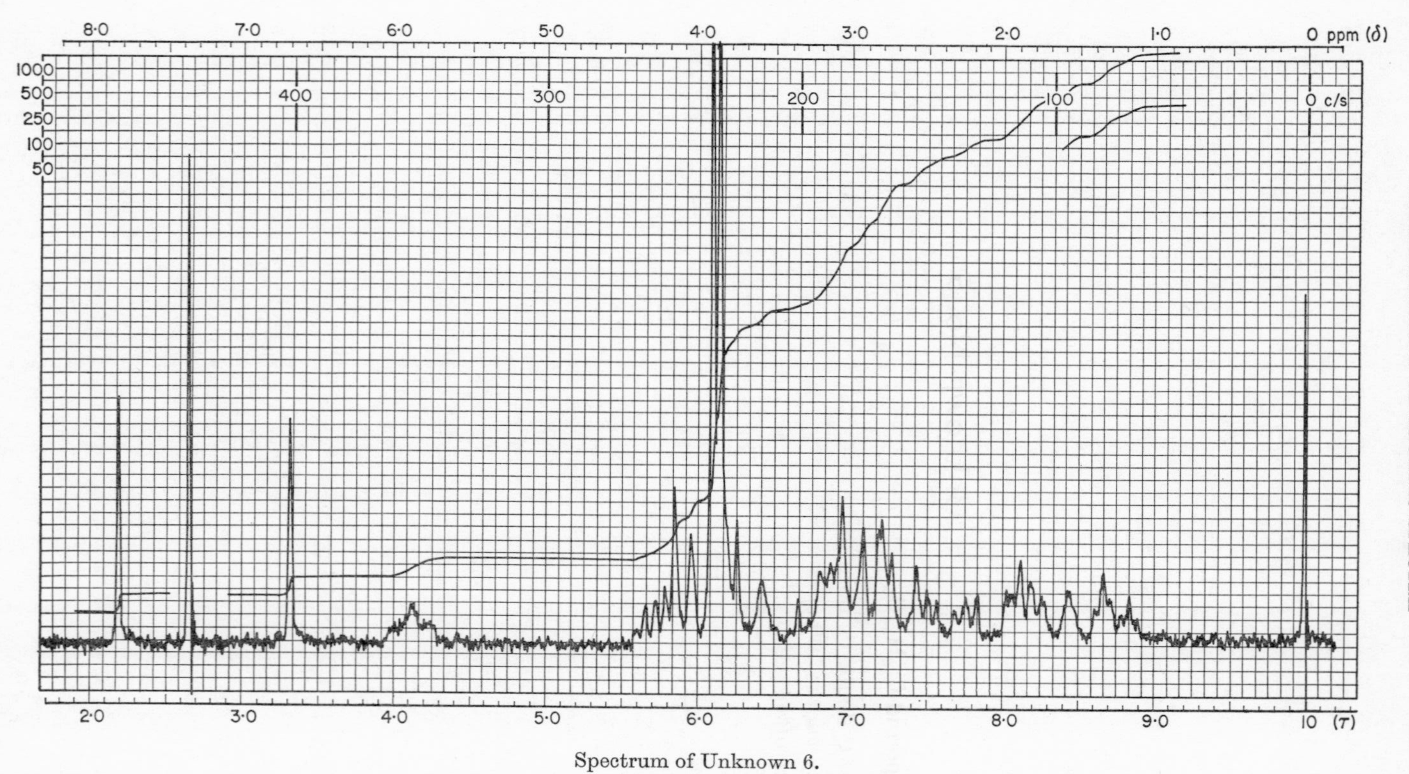

Spectrum of Unknown 6.

Unknown No. 6

Solvent: deuterochloroform
Sweep width: 500 c/s
Sweep offset: zero

This compound is brucine ($C_{23}H_{26}O_4N_2$): assign the signals in the spectrum.

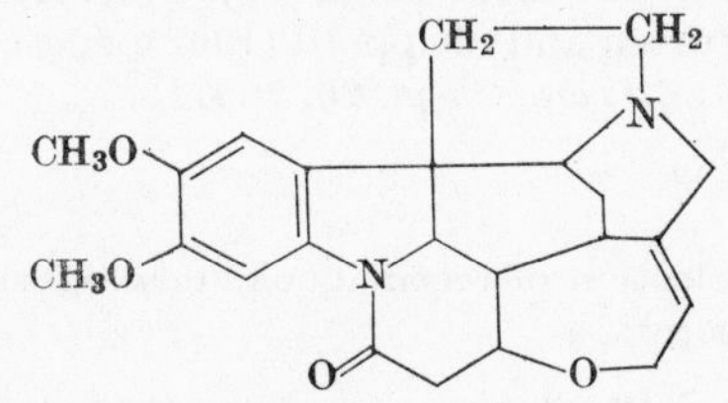

References

1a. Jackman (1959). "Applications of Nuclear Magnetic Resonance Spectroscopy in Organic Chemistry". Pergamon Press, London.
1b. Roberts (1959). "Nuclear Magnetic Resonance. Applications to Problems in Organic Chemistry". McGraw-Hill, New York.
Roberts (1961). *J. chem. Educ.* **37**, 581.
Conroy (1960). Nuclear Magnetic Resonance in Organic Structural Elucidation. *In* "Advances in Organic Chemistry", ed. by Raphael, Taylor and Wynberg. Vol. 2, p. 265. Interscience, New York.
2a. Varian N. M. R. Spectra Catalog, Varian Associates, Palo Alto, California, 1963.
2b. Chamberlain (1959). *Analyt. Chem.* **31**, 56.
2c. Meyer, Saika and Gutowsky (1953). *J. Amer. chem. Soc.* **75**, 4567.
2d. Banwell and Sheppard (1962). *Disc. Faraday Soc.* **34**, 115.
2e. Bernstein and Sheppard (1962). *J. chem. Phys.* **37**, 3012.
2f. Banwell and Sheppard (1960). *Mol. Phys.* **3**, 351.
2g. Karplus (1959). *J. chem. Phys.* **30**, 11.
3a. Pople, Schneider and Bernstein (1959). "High resolution Nuclear Magnetic Resonance". McGraw-Hill, New York.
3b. Roberts (1961). "An Introduction to the Analysis of Spin-Spin Splitting in High Resolution Nuclear Magnetic Resonance Spectra". Benjamin, New York.
3c. Corio (1960). *Chem. Rev.* **60**, 363.
4. MacGillavry, Kreuger and Eichhorn (1951). *Proc. Acad. Sci. Amst.* **B54**, 449.
5. Jones, Katritzky, Murrell and Sheppard (1962). *J. chem. Soc.* 2576.
6. Tiers and Coon (1961). *J. org. Chem.* **26**, 2097.

Bibliography

The following are a selection of recent books dealing with the organic applications of NMR Spectroscopy:

Pople, J. A., Schneider, W. G. and Bernstein, H. J. (1959). "High Resolution Nuclear Magnetic Resonance", McGraw-Hill, London.

Jackman, L. M. (1959). "Applications of Nuclear Magnetic Resonance Spectroscopy in Organic Chemistry", Pergamon Press, London.

Stothers, J. B. (1963). Applications of Nuclear Magnetic Resonance Spectroscopy. *In* "Elucidation of Structures by Physical and Chemical Methods", Part I, "Technique of Organic Chemistry", ed. by Weissberger, A. Vol. XI Interscience, New York.

White, R. F. M. (1963). Nuclear Magnetic Resonance Spectra. *In* "Physical Methods in Heterocyclic Chemistry", ed. by Katritzky, A. R. Part II. Academic Press, New York.

Conroy, H. (1963). Nuclear Magnetic Resonance in Organic Structural Determination. *In* "Advances in Organic Chemistry", ed. by Raphael, R. A., Taylor, E. C. and Weinberg, H. Vol. 2, p. 265, Interscience, New York.

Gillespie, R. J. and White R. F. M. (1962). Nuclear Magnetic Resonance and Stereochemistry". *In* "Progress in Stereochemistry," ed. by de la Mare, P. B. D. and W. Klyne Vol. 3., Butterworths, London.

Roberts, J. D. (1959). "Nuclear Magnetic Resonance Applications to Organic Chemistry", McGraw-Hill, New York.

Roberts, J. D. (1959). "An Introduction to the Analysis of Spin-Spin Splitting in High Resolution Nuclear Magnetic Resonance Spectra". Benjamin, New York.

Silverstein, R. M. and Bassler, G. C. (1963). "Spectrometric Identification of Organic Compounds". Wiley, New York.

Wiberg, K. B. and Nist, B. J. (1962). "Interpretation of NMR Spectra". Benjamin, New York. (This is a collection of computed spectra for the following cases: AB, AB_2, ABX and ABC, AB_3, A_2B_2, AB_4, A_2B_3.)

Infrared Spectroscopy

Introduction

With one exception, the infrared spectra have been recorded on small, low-resolution instruments with sodium chloride optics. Because of this, corrected frequencies for the main peaks have been inserted in the printed spectrum. The two reference books used in the interpretation of the spectra were:

L. J. Bellamy, *The Infra-red Spectra of Complex Molecules*. Methuen, London, 1958.

R. N. Jones and C. Sandorfy, The application of Infrared and Raman Spectrometry to the Elucidation of Molecular Structure, Vol. IX of *Technique of Organic Chemistry*. Interscience, London, 1956.

These are referred to throughout this section as "B." and "J. and S.", respectively.

Infrared Spectra and their Interpretation

Spectrum No. 1

This sample is a C_7 hydrocarbon. The I.R. bands at 1600, 1500, 728 and 695 cm^{-1} are due to an impurity which absorbs weakly in the near ultraviolet. Sample presentation: pure liquid at a path length of 0·025 mm.

Since the material is a hydrocarbon, it is useful to begin with C—H stretching frequencies near 3000 cm^{-1}. There are no bands above this frequency; three types of hydrocarbon may therefore be ruled out; aromatic compounds (C—H stretching frequency near 3030 cm^{-1}, B., p. 65), alkynes which show a strong band at 3300 cm^{-1} (B., p. 58) and alkenes where the C—H stretch lies between 3040 and 3010 cm^{-1} (B., p. 34).

The main band at 2900 cm^{-1} clearly arises from the C—H bonds in a saturated hydrocarbon, while the smaller peak at 2872 cm^{-1} on the side of the main band is due to the C—H symmetrical stretch in a methyl group, (B., p. 13). As this band is not particularly strong, it would suggest that only a small number of methyl groups are present (see below).

A word of caution must be given; on the tacit assumption that the wavenumber calibration is accurate, we have made certain deductions, but if these deductions are not supported by other parts of the spectrum, the calibration should be checked, particularly in the region 4000–2000 cm^{-1}. With sodium chloride optics, moreover, this region suffers from poor resolution.

The presence of only weak absorption from 1500 to 2500 cm^{-1} confirms the absence of acetylene or olefin C—H bonds. In this region, the stretching frequencies of C≡C and C═C arise (B., p. 34), but where symmetrically substituted molecules of these types occur, as in dimethylacetylene or tetramethylethylene, there is no dipole moment change on stretching, and therefore no absorption band in the infrared, although a line does appear in the Raman spectrum. Even where the symmetry is modified with alkyl groups of different size (e.g. ethyltrimethylethylene), the C═C stretching band is often too weak to see.

A strong band at 1450 cm^{-1} arises from the methylene scissoring vibration, and is very strong in all aliphatic hydrocarbons (J. and S., p. 343); at 1460 cm^{-1} the antisymmetrical deformation of a methyl group is also to be found, but owing to the strong overlapping of the methylene

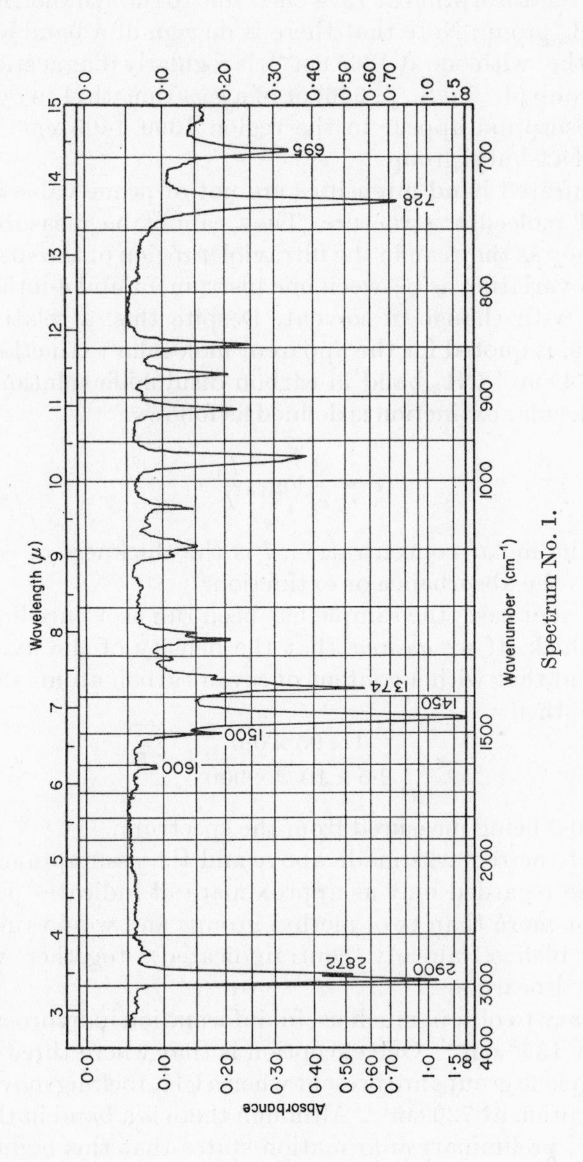

Spectrum No. 1.

scissoring vibration, obviously no conclusions as to methyl groups can be drawn from this frequency. Greater diagnostic value, however, can be attached to the absorption at 1374 cm^{-1} due to the symmetrical deformation of a CH_3 group. Note that there is no sign of a band at 1385 cm^{-1} which, together with one at 1365 cm^{-1}, is regularly diagnostic either of an isopropyl group (J. and S., p. 356) or of a *gem*-dimethyl in cyclohexanes: nor does absorption appear in the region 1390–1406 cm^{-1}, this being indicative of a t-butyl group.

As yet, infrared band intensities are not of prime value in the determination of molecular structure. They cannot be measured with the same accuracy as they can in the ultraviolet region of the spectrum: they are liable to variation as between one instrument and another, and they can change with change of solvent. Despite this, a relatively narrow range, 33–56, is quoted for the apparent molecular extinction coefficient ϵ of the 1374 cm^{-1} CH_3 band in carbon disulphide solution (J. and S., p. 347). Molecular extinction is defined as follows:

$$\epsilon = \frac{1}{cl}\log_{10}\frac{I_0}{I}$$

where c is the molar concentration, l is the thickness of cell (cm) and $\log_{10}(I_0/I)$ is the absorbance or extinction.

In the present case, the sample has been run as a pure liquid in a cell 0·025 mm thick. If we *assume* that the density of the hydrocarbon is about 0·9 and that with a content of seven carbon atoms the molecular weight is 95, then

$$\epsilon = \frac{1 \times 95 \times 0{\cdot}6}{2{\cdot}5 \times 10^{-3} \times 900} = 25$$

$\log(I_0/I) = 0{\cdot}6$ being measured from the spectrum.

In view of the remarks made above and the assumptions made, this figure can be regarded only as approximate; it indicates probably one, certainly not more than two, methyl groups and would rule out tetra-substituted olefins (already contraindicated), together with highly branched hydrocarbons.

It is not easy to obtain much useful information in hydrocarbons from bands below 1350 cm^{-1}. One exception is that where three or more adjacent methylene groups are present when a CH_2 rocking movement gives rise to absorption at 720 cm^{-1}. Although there is a band in the spectrum at 728 cm^{-1}, preliminary information states that this is due to an impurity. Note the sharpness of the bands in this region. Flexible molecules have many rotational isomers each of which has slightly different skeletal vibrations and in consequence only very broad bands are to be expected (band width at half height of about 10 cm^{-1}). It is useful to compare the appearance of typical spectra from molecules with long methylene chains (J. and S., p. 345).

From all considerations so far, a C_7 saturated cyclic hydrocarbon can most usefully be considered. The methyl band at 1374 cm^{-1} rules out a cycloheptane and, if attention is confined to five- and six-membered rings, seven compounds remain: ethylcyclopentane, dimethylcyclopentane (five isomers) and methylcyclohexane.

At this point, the problem would normally involve comparing the unknown, with authentic spectra of these seven possibilities but, in this case, additional help is available in the impurity stated to be present.

For this, the band at 1500 cm^{-1} is characteristic of the symmetrical infra-red-active ring vibration in aromatic structures (B., p. 72) often accompanied by a second band at 1600 cm^{-1}: both are rather weak in the present example. The strong absorption at 728 cm^{-1}/cm is due to the out-of-plane hydrogen deformation in a mono-substituted benzene. The range given for this nearly overlaps with *ortho* disubstituted benzenes, but the second band at 695 cm^{-1} occurs only in mono-substituted, *meta*- or 1,3,5-trisubstituted benzenes. Together with the absorption in the ultra-violet region of the spectrum, this strongly suggests toluene as the impurity. Methylcyclohexane is, of course, made by the hydrogenation of toluene, and complete removal of any unchanged starting material is not easily achieved.

These facts identify the unknown as methylcyclohexane and the impurity as toluene.

Spectrum No. 2

This sample is a liquid hydrocarbon of molecular formula $C_{13}H_{24}$. Sample presentation: capillary film between rock salt plates.

An initial glance at the spectrum shows that there are few bands present, and thus we are dealing with a relatively simple molecule.

Working from the high wavenumber end of the spectrum, the band at 3304 cm^{-1} must be due to a C—H stretching mode, because of the known absence of oxygen and nitrogen. It is not likely to be an overtone of a fundamental mode, since there are no bands apparent in the 1650 cm^{-1} region ($\frac{3304}{2}$ cm^{-1}), and its position at 3304 cm^{-1} indicates an acetylenic C—H group (B., p. 58).

The bands at 2923 and 2854 cm^{-1} occur in the region of saturated C—H stretching modes, and the exact positions indicate the antisymmetrical and symmetrical stretches, respectively, for methylene groups. As these bands are the most intense in the whole spectrum, one may infer the presence of a large number of methylene groups. It should be noted that there is no clear sign of bands or inflections around 2960 or 2872 cm^{-1}, which would indicate a CH_3 group.

The band at 2128 cm^{-1} occurs in the triple bond stretching region and, in the absence of nitrogen, can be assigned only to —C≡C— (B., p. 58) Its exact position and the occurrence of the band at 3304 cm^{-1} discussed above, indicate that we have a mono-substituted acetylene. The absence of any bands in the 1500–1700 cm^{-1} region suggests that there are no carbon-carbon double bonds.

Some of the bands in the C—H deformation region (1300–1500 cm^{-1}) provide useful information. An intense peak at 1467 cm^{-1} probably arises from the $-CH_2-$ groups previously noted: the inflection at 1435 cm^{-1} could be the antisymmetrical mode of a CH_3 group (cf. Spectrum No. 1), an assignment strengthened by the appearance of a band at 1378 cm^{-1} which could arise from the symmetrical mode. The fact that the methyl groups shows up in the 1400 cm^{-1} region and not in the 3000 cm^{-1} is perhaps not surprising in this particular spectrum which was run on an instrument using a rock salt prism. This has relatively poor resolution in the 3000 cm^{-1} region, so that a band due to a methyl group would not necessarily be resolved when a number of methylene groups was also present.

Since we know that a mono-substituted acetylene is present, the band at 1239 cm^{-1} might be a combination band which has been reported in this region for a number of other mono-acetylenic compounds (B., p. 61).

The more intense band at 719 cm^{-1} probably arises from a rocking

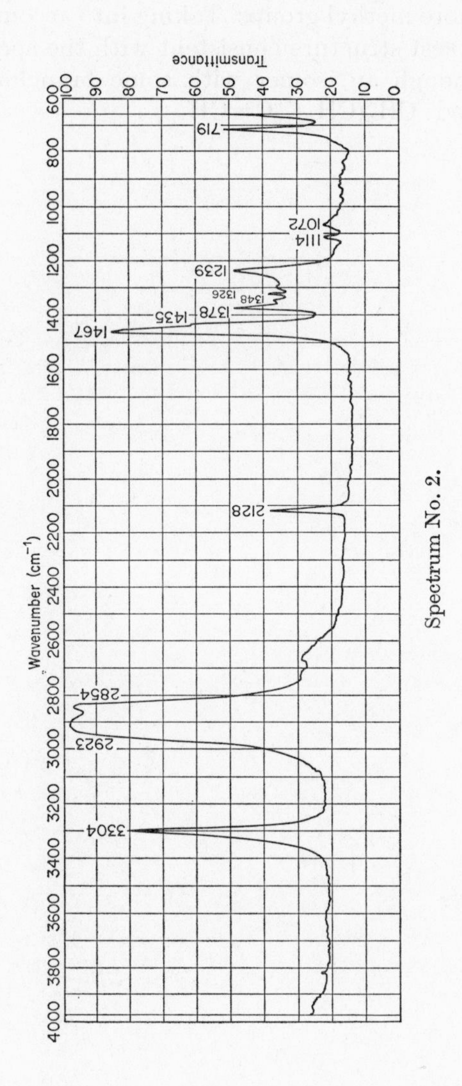

Spectrum No. 2.

mode of the methylene groups, as it is known that compounds with four or more adjacent methylene groups show a band near 720 cm^{-1} (B., p. 27).

In summary, the infrared spectrum of this compound indicates a terminal acetylenic group, at least four adjacent methylene groups, and one or possibly more methyl groups. Taking into account the molecular formula, the simplest structure consistent with the spectrum is that of tridec-1-yne, although an isomer with some branching in the chain cannot be excluded: $CH_3(CH_2)_{10}C{\equiv}CH$.

Spectrum No. 3

This sample was obtained from a solvent extract of a tar. It is a colourless solid, neutral in character and absorbs in the near ultraviolet; the elemental analysis indicates that only carbon and hydrogen are present. Sample presentation: (A) solution in carbon tetrachloride, (B) solution in carbon disulphide. Path length 0·1 mm., concentration 12%.

The X—H stretching region shows that no O—H or N—H groups are present. This is consistent with the neutral character of the material and with the elementary analysis.

Acetylenic C—H is also excluded ($\sim$3300 cm^{-1}). The clearly defined absorption at 3040 cm^{-1} is associated unambiguously with a C—H stretching vibration and, since it lies above 3000 cm^{-1}, it can be assigned to =C—H either in an olefin or in an aromatic compound. The fact that the material was obtained from coal tar suggests an aromatic structure, although its does not exclude the possibility of an olefin. Absorption in the near ultraviolet is consistent with an aromatic chromophore; for olefins to absorb at wavelengths longer than $\sim$205 mμ, a conjugated double-bond system is necessary.

The absence of absorption in the region 3000–2800 cm^{-1} shows that none of the hydrogen in the compound is attached to saturated carbon; alkyl side chains or hydroaromatic rings are thus eliminated. This severely limits the possibilities and emphasizes the importance of a careful analysis of the X—H stretching region.

Examination of the spectrum down to 1600 cm^{-1} further limits the choice of possible structures. There is no evidence for C≡N or C≡C; these have stretching vibrations between 2300 and 2100 cm^{-1} and, with the possible exception of *symmetrically substituted* structures discussed later, would give rise to infrared absorption. Further, absorption bands between 1900 and 1600 cm^{-1} are of but low intensity, which shows that, if C=O groups are present at all, their concentration is low.

In the C=X stretching region it is important to look for evidence of C=C absorption, as hydrogen attached to unsaturated carbon has already been detected. While the absence of absorption between 1680 and 1600 cm^{-1} does not completely eliminate olefinic structures, it so limits their type that the emphasis must be placed strongly in terms of an aromatic compound. This emphasis is increased by the presence of a group of absorption bands between 1600 and 1430 cm^{-1}, namely the two peaks at 1595 and 1568 cm^{-1} and the strong band at 1482 cm^{-1}. These lie in the region where skeletal vibrations in aromatic structures give rise to characteristic absorption (J. and S., pp. 394–398).

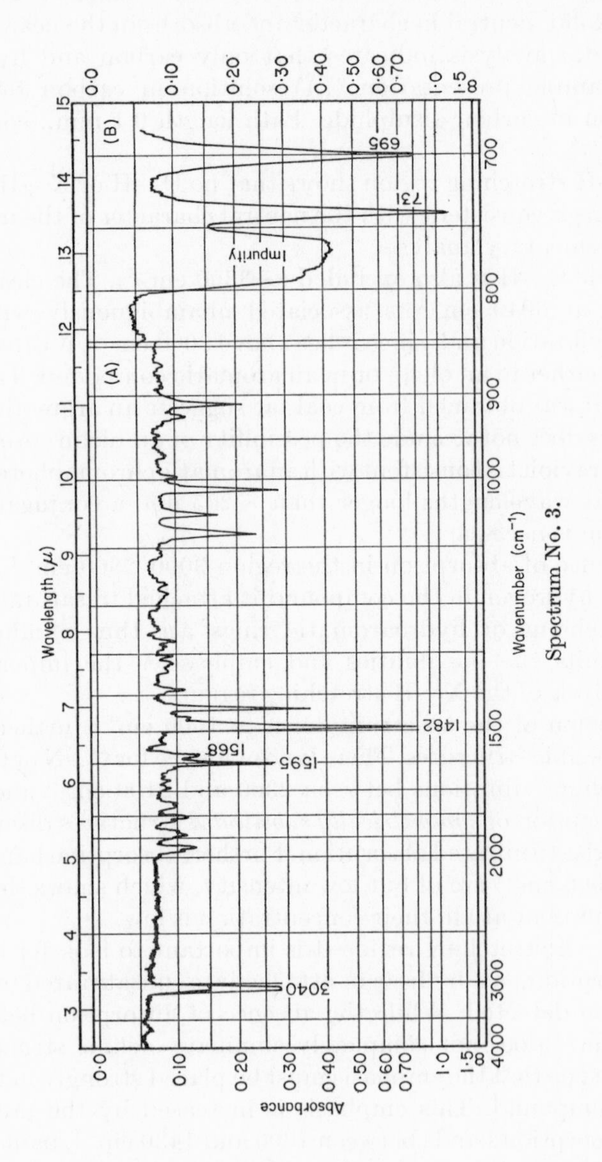

Spectrum No. 3.

Most of the region between 1400 cm^{-1} and 650 cm^{-1} is populated by absorption bands of medium intensity and, in common with the great majority of infrared spectra, needs caution in interpretation. Any "group" frequencies in this region will always occur with those of a less specific character and, unless they are associated with bands of high intensity, misleading conclusions can be drawn. In the present case, there are two strong absorption bands at 731 and 695 cm^{-1} which can be interpreted without difficulty. Absorption in this region is associated with out-of-plane vibrations of aromatic C—H groups, and can provide an important clue to the pattern of ring substitution. In simple benzene ring systems, the only type showing absorption both at 695 and at 730 cm^{-1} is a monosubstituted benzene.

As much information as can be extracted reasonably from the spectrum has now been obtained, and the simplest structure consistent with this information is a mono-substituted benzene. The substituent is not a *saturated* chain or ring structure and therefore the obvious alternative is a second phenyl group to give diphenyl.

Although diphenyl is the simplest compound consistent with the spectrum, it is worthwhile to consider further possibilities. These illustrate other aspects of the analysis which may mislead in certain restricted cases. One such example is a structure in which two aromatic rings are linked by a C═C or C≡C group, so that the molecule has a centrosymmetric configuration. In these, however, the vibration associated primarily with the stretching of the linking group would not give rise to a change in the molecular dipole moment and, since this is a necessary condition for absorption in the infrared, no bands associated with olefinic or acetylenic linkages would be observed: tetraphenylethylene and diphenylacetylene are examples of such structures. *trans*-Stilbene also is a compound where the carbon–carbon double bond stretching vibration is inactive. In this case, however, one would expect absorption associated with out-of-plane CH deformation vibrations (B., p. 46); no such absorption is present in this spectrum.

At this stage, only a direct comparison with the spectra of known structures can decide between the alternatives discussed. In fact the spectrum is that of diphenyl. Simple physical properties must never be ignored: they may be of considerable help in narrowing the range of possibilities. Where information given with the sample is meagre, the number of structures consistent with a group frequency analysis is likely to be considerably greater than in the present example. It is the exception rather than the rule for a complete structural identification to be made from the infrared spectrum alone. The presence of absorption associated with the stretching of C—H bonds where hydrogen is attached to a saturated carbon atom would have greatly extended the range of possible structures in the preceding discussion. Had no information been given with the sample, then elements other than carbon and hydrogen would perforce have been considered. Certain substituents containing oxygen or nitrogen were easily eliminated in the foregoing analysis, but C_6H_5X, (where X is –N═N–, NO_2, Cl, Br, or I) could not have been excluded. This underlies the fact that basic information, such as an analysis for elements, can be of crucial importance as a background to the interpretation.

Three main spectral regions have been used in the analysis of aromatic structures: these are, near 3050 cm^{-1} for the C—H stretching absorption, the 1600–1500 cm^{-1} group, and the 700–900 cm^{-1} range where patterns associated with ring substitution occur. If hydrogen is directly attached to the aromatic rings, absorption should occur in each of the above three regions, whatever the type of aromatic compound. The correlations published for the 700–900 cm^{-1} patterns have been based mainly on single ring compounds with a bias towards hydrocarbons. Some correlation tables show wider ranges than others for the bands in this group, presumably because they have been derived from a more diverse selection of structures. It is advisable to be cautious in the interpretation of the 700–900 cm^{-1} region if there is evidence from other parts of the spectrum that groupings of a non-hydrocarbon character are present; this applies particularly if the degree of substitution is high. In general, it is best to consider only the strong absorption bands in the region and to display the spectrum at a path length such that the relative intensities of bands within the group are apparent.

Between 2000 and 1650 cm^{-1}, patterns of weak absorption bands characteristic of the type of substitution have been found for a number of simple ring aromatic compounds. Some low intensity bands can be seen in this region in the diphenyl spectrum. With a greater path length of sample the form of such patterns can be useful, particularly in hydrocarbons where interfering absorption is unlikely to occur, (J. and S., pp. 397–399).

The overall picture of the relative intensities of absorption bands in the three main groups characterizing aromatic compounds is one of high intensity in the 700–900 cm^{-1} group and a variable intensity in the 1600–1500 cm^{-1} group, but in many cases of simple single-ring compounds there are strong bands near 1500 cm^{-1}; finally, a low intrinsic intensity for the C—H stretching absorption. In hydrocarbons this relationship means that the first two groups stand out from the remainder of the spectrum, and the diphenyl pattern is an illustration of this. It is a feature which can be used as a rough guide to distinguish hydrocarbons from other structures in the aromatic series. In complex systems, it frequently happens that aromatic rings may form only a minor part of the whole structure and may then be difficult to characterize. This is particularly true of the C—H stretching absorption which can be obscured by strong neighbouring absorption from C—H groups in saturated systems or by the broad absorption from O—H or N—H groups in hydrogen-bonded structures.

Spectrum No. 4

This material has a melting point of 121–123°, a pK_a of ~10 and it absorbs in the near ultraviolet. The sample was pressed in a potassium bromide disk (~1 mg/400 mgKBr) and the spectrum was run with a blank disk in the reference beam.

The X—H stretching region reveals at the outset one major feature, a strong absorption band at 3300 cm^{-1} associated with an O—H or N—H stretching vibration in a system which has polymeric association through strong intermolecular hydrogen-bond interaction. In this example, the information provided favours a weakly acidic OH rather than N—H. Without the pK value it would be difficult to decide between these possibilities, although in many cases an NH_2 group shows two bands in this region and these are narrower than the OH absorption. For materials associated by *inter*molecular hydrogen bonds, it is helpful to examine the spectrum at greater dilutions. Thereby the position of the unbonded OH stretching absorption can aid in resolving difficulties.

The very broad X—H stretching vibration is an important feature for all strongly hydrogen-bonded groups. In this region, it distinguishes clearly between this situation and the alternative assignment (based on position only) of an acetylenic C—H. At first sight, the band on this trace does not appear to be strikingly broader than some others in the spectrum, but it should be remembered that the wavenumber scale 3000–4000 cm^{-1} is greatly compressed into a narrow width of chart paper; this is consequent on the instrument used recording linearly in wavelength rather than frequency. Thus the width of the band is at least 300 cm^{-1}. This should be borne in mind when spectra obtained from different instruments are compared; the appearance of wavenumber scales can differ widely.

The weak absorption band at approximately 3050 cm^{-1} which occurs as a shoulder on the O—H absorption can be assigned, as in the previous example, to an olefinic or aromatic C—H stretching frequency. The fact that the material absorbs in the near ultraviolet favours an aromatic structure, and the analysis so far is entirely consistent with a phenol. Moreover, there is no detectable absorption between 3000 and 2800 cm^{-1} from C—H groups in saturated structures, and we may eliminate alkyl side chains or hydroaromatic rings.

The absence of significant absorption down to 1635 cm^{-1} eliminates C≡N, C≡C and C=O, with the same reservations for acetylenic C≡C as discussed in the previous example.

The group of bands in the region of 1600 cm^{-1} and 1500 cm^{-1} leaves little doubt that the substance is aromatic. All the bands in the 1600 cm^{-1}

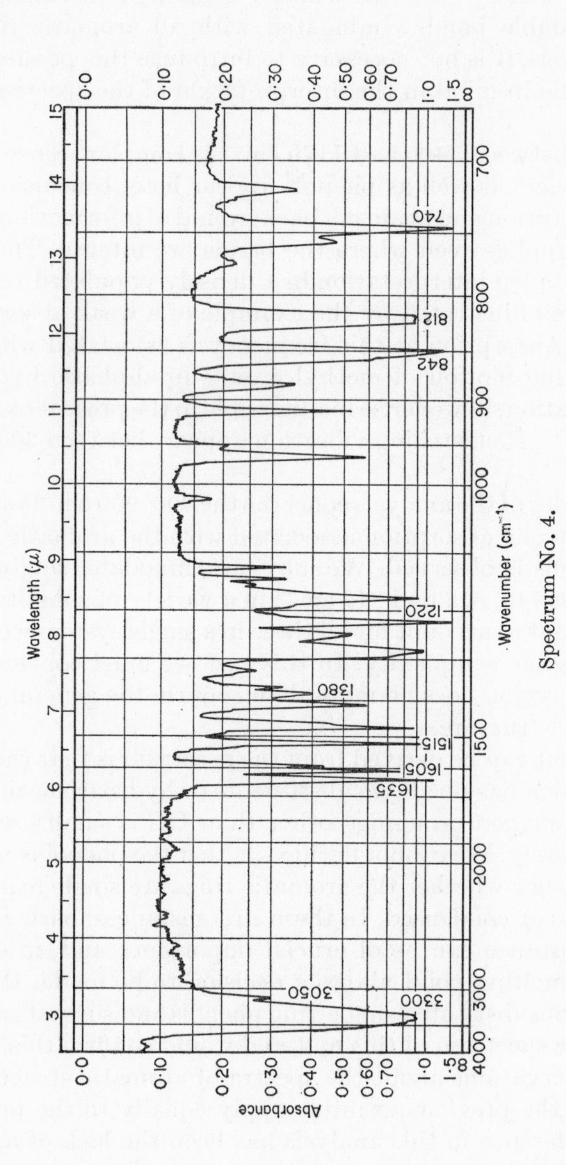

Spectrum No. 4.

group can be accommodated without diffculty in the range of characteristic skeletal stretching modes, but one of them, at 1635 cm^{-1}, overlaps the region covered by C=C stretching vibration of olefins, and in particular, with double bonds conjugated with an aromatic ring. In the simplest analysis, it is not necessary to introduce this possibility, but it should be borne in mind in the interpretation of the spectrum at lower frequencies.

The region between 1400 and 1170 cm^{-1} is complex. Since absorption due to aryl ethers as well as phenols appear here, conclusions must be derived with caution and against a background of information from other regions; this applies even where the bands are intense. The danger of attempting detailed interpretation in a densely populated region of the spectrum is well illustrated by the example of a weak absorption band at 1380 cm^{-1} Absorption at this frequency is associated with the symmetrical bending motion of methyl groups in aliphatic hydrocarbons. This interpretation, however, is inadmissible in the present example since no saturated C—H stretching vibration appears between 3000 and 2800 cm^{-1}.

The remainder of the analysis concerns the 700–900 cm^{-1} region where, as expected, strong absorption associated with the aromatic CH out-of-plane vibrations is observed. We may accommodate, singly, the three major bands at 740, 812 and 842 cm^{-1} in a variety of substituted single-ring systems, but to account for all three in a single system would require structures of some complexity. In this case we must conclude that the 700–900 cm^{-1} region does no more than confirm the general assignment to an aromatic structure.

The most that can be derived from the spectrum is that the substance is a phenol which has no alkyl substituents or hydroaromatic rings. The possibility of oxygen groupings other than OH is small but cannot be eliminated entirely. We cannot decide whether the phenol is monohydric or polyhydric, nor whether the aromatic rings are single (e.g. in a polyphenyl system) or condensed. In these circumstances, basic information about the substance can be of crucial importance and, in our present example, the melting point allows a decision to be made. It eliminates all the simple unsubstituted single-ring phenols and suggests β-naphthol. A check on the spectrum of this material would confirm this conclusion.

General observations about the spectra of aromatic structures which were made in the previous example apply equally to the present case. The major difference in this analysis has been the lack of any detailed interpretation of the 700–900 cm^{-1} region. One reason for this difficulty is the condensed ring structure. The complexity of the patterns in this region for condensed ring compounds can be such that unambiguous interpretation is impossible. Some progress has been made in the analysis of the region for condensed hydrocarbons by considering each ring as a separate substituted benzene with each of the shared ring positions counted as a substituent. However convenient this form of analysis may be in certain cases, it cannot be generalized, and an attempt to apply the argument to the present example would have led to incorrect conclusions.

Spectrum No. 5

The crude acid fraction from a sample of Oriental Storax resin was esterified (ethanol–hydrochloric acid) and subjected to preparative gas–liquid chromatography. The major component was isolated as a liquid, b.p. 271°/760 mm. Sample presentation: capillary film between rock salt plates.

The history of the sample provides the minimum of assistance; thus the substance is reputedly an ester, probably free from contamination with free acid or alcohol, but possibly containing traces of closely related esters or of the chromatographic stationary phase.

It is important first to confirm that the substance is indeed an ester. The spectrum, for a liquid film, shows a strong band in the carbonyl region at 1715 cm^{-1}. This frequency is low for a saturated ester but is in the correct position for the ester of an α,β-unsaturated carboxylic acid. The band at 1173 cm^{-1} can probably be assigned to the C—O stretching frequency of this ester group.

Apart from some very weak bands near 3400 and 3600 cm^{-1}, which may well be overtone bands, the only absorption in the single bond stretching region is close to 3000 cm^{-1} in that part of the spectrum usually assigned to carbon–hydrogen stretching frequencies. Two bands at $\sim$2960 and $\sim$2880 cm^{1} suggest the presence of saturated carbon–hydrogen groupings and a third band at $\sim$3050 cm^{-1} possibly indicates an unsaturated carbon–hydrogen grouping. The presence of saturated hydrocarbon groups is required by the ethyl ester function and, while in itself this adds little to our analysis, the intensity and contour of this doublet suggests that few additional saturated hydrocarbon groups can be present. The band at 3050 cm^{-1} lends support to the formulation of the carbonyl band as that of an α,β-unsaturated ester.

In the carbonyl stretching region, in addition to the strong band at 1715 cm^{-1}, there is a band of medium intensity at $\sim$1640 cm^{-1}. Such a band, which may be ascribed tentatively to the stretching frequency of a carbon–carbon double bond, is required by the presence of an α,β-unsaturated ester. The intensity of this band is greater than would be expected for an isolated olefinic system, and provides further support for this assignment (B., p. 41; J. and S., p. 374). Now out-of-plane deformation frequencies of olefinic carbon–hydrogen bonds provide useful, although not entirely reliable, confirmation of the existence of such structural units. Since the ester must be at least a di-substituted olefin we would expect such deformation bands to be strong and in the region 800–1000 cm^{-1}. Although this part of the spectrum contains a number of weak bands there is also a strong band at 980 cm^{-1} where the deformation

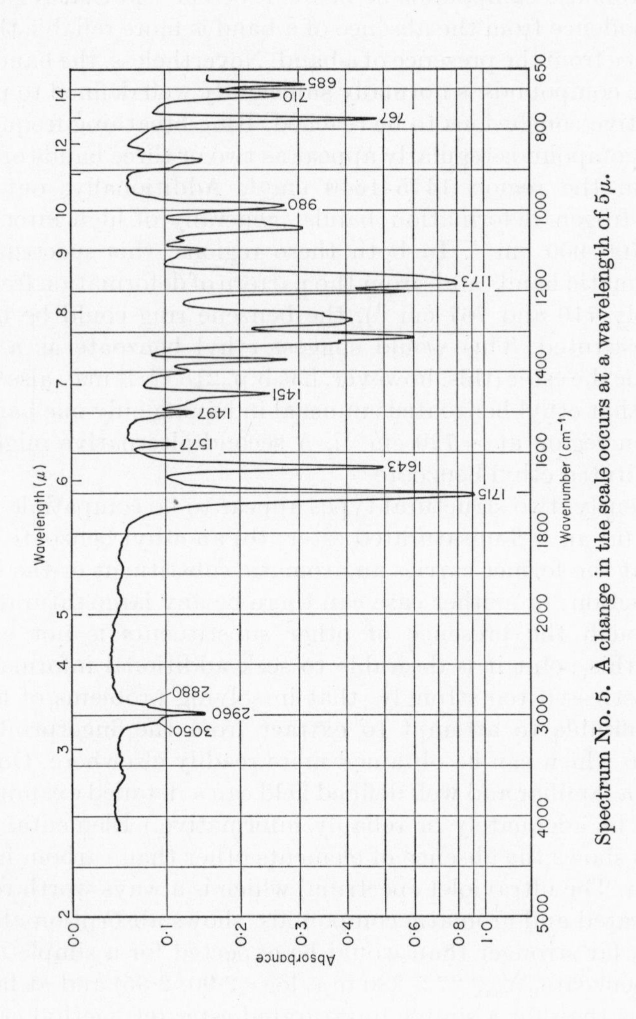

Spectrum No. 5. A change in the scale occurs at a wavelength of 5μ.

frequency of *trans*-1,2-di-substituted olefins is known to occur (B., p. 45).

Such evidence, however, does not exclude the possibility that the α,β-unsaturation is due to an aryl group, since the band at 1640 cm^{-1} could be assigned to an aromatic "ring breathing" frequency. The characteristic bands of aromatic compounds lie below 1600 cm^{-1} in that region where negative evidence from the absence of a band is more reliable than positive evidence from the presence of a band. Nevertheless, the band pattern of aromatic compounds is normally sufficiently well defined to permit at least tentative conclusions to be reached. Ring breathing frequencies of benzenoid compounds regularly appear as two or three bands of medium intensity in the region 1475–1640 cm^{-1}. Additionally, out-of-plane carbon–hydrogen deformation bands, generally of high intensity, are found at 700–900 cm^{-1}. In both these regions, this spectrum shows typical aromatic bands and, from the pattern of deformation frequencies (i.e. possibly 710 and 767 cm^{-1}), the benzene ring could be mono- or *ortho*-disubstituted. This would suggest ethyl benzoate as a possible structure for the ester (this, however, has b.p. 215°C; it may also be noted in passing that ethyl benzoate is unusual in having only one band in the deformation region at $\sim$720 cm^{-1}). A second alternative might be an *ortho*-substituted ethyl benzoate.

Consequently, two structural types appear to be compatible with the spectrum: (a) an α,β-unsaturated ester; (b) an ethyl benzoate. It may well be that the former carries an aromatic substituent or the latter an olefinic function; in neither case can there be any large saturated alkyl units although the presence of other substituents is not excluded. Clearly at this point it is desirable to seek additional information. (It cannot be stressed too strongly, that in solving problems of this sort, it is unprofitable to attempt to extract from the fingerprint region, information which can be obtained more readily elsewhere. Only when working in a familiar and well defined field can a detailed examination of this region be adequately or reliably informative.) Elemental analysis ($C_{11}H_{12}O_2$) shows the absence of elements other than carbon, hydrogen and oxygen. The ultraviolet spectrum, which is always worth recording for unsaturated and aromatic compounds, shows absorption at 277 mμ (log ϵ 4·30) far stronger than would be expected for a simple benzoate (cf. ethyl benzoate, λ_{max} 272, 280 mμ, log ϵ 2·90, 2·85) and at far longer wavelengths than for a simple unsaturated ester (cf. methyl crotonate, λ_{max} 212, 250 mμ, log ϵ 4·16, 2·13). The obvious implication of these results is that the benzenoid ring and olefinic system are both conjugated with the carboxyl group, as for example in ethyl cinnamate (C_6H_5—CH═CH—$COOC_2H_5$). Such a substance would be expected to show in its infrared spectrum the characteristics of both an α,β-unsaturated ester and an aromatic ring. Comparison of the spectrum of authentic ethyl cinnamate with the spectrum under discussion will show that they are identical. In making such comparisons, it should be remembered that all spectra must be run under identical conditions.

Spectrum No. 6

This sample was isolated in good yield from the pyrolysis of isopropenyl acetate ($CH_2{=}C(CH_3){-}OOCCH_3$). It is a colourless liquid, b.p. 133–4°C. which gave a single peak when examined by gas chromatography. Sample presentation: capillary film between rock salt plates.

From the history of the sample one may deduce that reasonable structural units to expect are *C*-methyl groups (probably in an isopropyl group), oxygen functions (perhaps as hydroxyl, carbonyl or ester groups) and unsaturated systems. It can be established that the spectrum is not that of the starting material (b.p. 96°C), as there is no band in the carbonyl region at a frequency corresponding to that of a vinyl ester (~ 1770 cm^{-1}); ideally, this conclusion should be reached only after the direct comparison of the two spectra.

In the single bond stretching region, only one broad band is clearly indicated with a centre near 2940 cm^{-1}; this band is in the position expected for the stretching frequencies of carbon–hydrogen bonds on saturated carbon. The spectrum apparently contains no bands indicative of free hydroxyl or of carbon–hydrogen bonds on unsaturated carbon.

In the carbonyl region there are three bands, two of relatively low intensity at 1709 and 1727 cm^{-1} and one of high intensity at 1613 cm^{-1}. The reliability of the carbonyl region in making structural assignments derives from two factors; firstly, the freedom of this part of the spectrum from other absorptions and, secondly, the fact that carbonyl bands are of high intensity ($\epsilon \sim 200$–800). Consequently, the low intensity of the higher frequency bands prevents their simple assignment to carbonyl groups. In this situation, three obvious possibilities must be considered: (a) the bands are not derived from carbonyl stretching frequencies at all but are due, for example, to overtone or combination bands; (b) they are carbonyl bands of exceptionally low intensity, a situation which must imply some structural peculiarity; (c) the sample is a mixture, at least one minor component of which contains normal carbonyl groups. Of these, the third possibility represents the most familiar cause of weak bands in the carbonyl region. If these bands are due to carbonyl-containing impurities, then their position suggests they be assigned to ketone or α,β-unsaturated ester carbonyl groups. However, the sample distils and behaves chromatographically as a pure substance and so the plausibility of this explanation, although not entirely eliminated, is reduced. If either (a) or (b) represents the true explanation of these bands, no useful assignment can be made at this stage.

The third band, at 1613 cm^{-1}, is at a frequency close to those expected for olefinic and aromatic carbon–carbon stretching frequencies. The

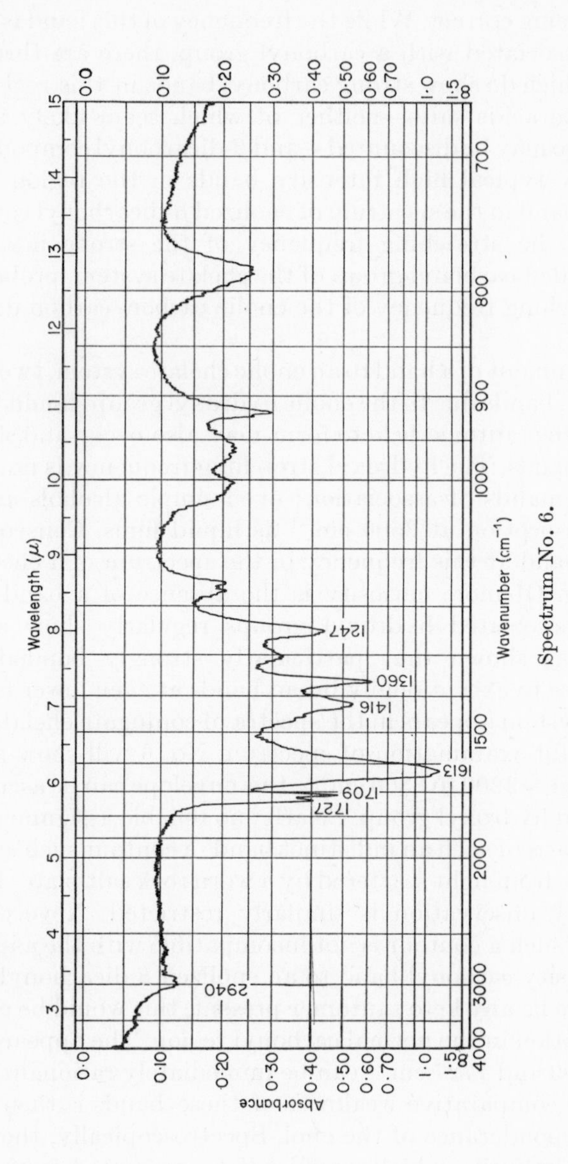

Spectrum No. 6.

absence of other bands in the region 1450–1600 cm^{-1}, however, suggests that it is not due to an aromatic ring stretching frequency and its extremely high intensity further diminishes the likelihood of either of these assignments being correct. While the frequency of this band is lower than is normally associated with a carbonyl group, there are three types of compounds which do show strong carbonyl bands in this region: amides and carboxylic acids salts—neither of which seem likely as reaction products and so may be discounted—and β-dicarbonyl compounds, which as enols, show typical high intensity bands in the region 1550–1640 cm^{-1}. Such a band in the spectrum of enolized β-dicarbonyl compounds is attributed to the stretching frequency of the strong intramolecular hydrogen-bonded carbonyl group of the chelate system, probably mixed with the stretching frequency of the enolic carbon–carbon double bond (B., p. 142).

By the assignment of a band to an enolic chelate system, two corollaries follow. Firstly, bands due to the enolic hydroxyl group should be present; and secondly the tautomeric keto form may also occur and should show recognizable bands. The hydroxyl stretching frequency is notably sensitive to the demands of association: even simple alcohols and phenols show little absorption at 3500 cm^{-1} as liquid films. Consequently the absence of a band at this frequency in the spectrum of a chelate enol is not surprising. Of more curiosity is the absence of a band near 3200 cm^{-1} where associated hydroxyl groups regularly show absorption. Experience has shown that particularly strongly bonded hydroxyl groups give rise to exceptionally broad bands at even lower frequencies. Such a band system appears in the spectra of conjugate chelate hydroxyl systems. Careful examination of spectrum No. 6 will show a flat band stretching from $\sim$3200 to 2300 cm^{-1}; this envelope can be ascribed to the chelate bonded hydroxyl group. Clearly, no reliable assignment could be made on the basis of such an indistinct band; a contour such as this could quite well arise from light scattered by a worn rock salt plate. Its value as a confirmatory observation is similarly restricted. Nevertheless, the appearance of such a contour is not incompatible with the assignment of the high intensity carbonyl band to an enolized β-dicarbonyl system.

Should there be any keto tautomer present, this would be expected to exhibit absorption in the normal carbonyl region. The appearance of the doublet at 1709 and 1727 cm^{-1} can be immediately rationalized in these terms and the comparative weakness of these bands is then seen to be due to the preponderance of the enol. Spectroscopically, the keto form represents an impurity which would not be separated from the major component by simple purification techniques.

At this stage it is possible to summarize the spectroscopic evidence as consistent with the sample being a tautomeric mixture of enol and keto forms of a β-dicarbonyl derivative. Although it may be intellectually attractive, it is uneconomic to pursue the analysis of the spectrum beyond this point if the primary objective is the identification of the pyrolysis product. Chemical tests ($FeCl_3$, $CuCl_2$, action of alkali) would confirm a β-dicarbonyl system and elemental analysis would allow restrictions to be placed on the possible chemical structures ($C_5H_8O_2$). Chemical intuition suggests that a reasonable re-arrangement of isopropenyl acetate could lead to acetylacetone; comparison of the spectrum with that of an authentic sample would establish this as the identity of the pyrolysis product.

Spectrum No. 7

This sample has a molecular formula, $C_{12}H_{13}N$. Sample presentation: potassium bromide disk (1 mg/200 mg KBr).

The spectrum is fairly complex, especially below 1500 cm^{-1} and contains about forty prominent bands, so that one cannot expect to assign more than a small number of them. The sharpness of these bands probably indicates an aromatic compound.

At the high frequency end of this spectrum lies the second strongest band in the whole spectrum, at 3420 cm^{-1} and, in the absence of oxygen, this can be safey assigned to an N—H group (B., p. 251). Since there is only one and not two bands in this region, an NH_2 group is ruled out. The high intensity of this band should be noted; it is known that the intensity of N—H bands increases with increasing acidity of this hydrogen atom, being particularly high in compounds such as pyrrole.

The weak band at 3080 cm^{-1} suggests an aromatic C—H stretching frequency. The pair of bands at 2930 and 2850 cm^{-1} are typical of those shown by methylene groups. There is no sign either of a band or of an inflection around 2960 cm^{-1} which would have indicated the presence of a methyl group (see below).

The pattern of weak absorption bands between 1650 and 2000 cm^{-1} is characteristic of partly substituted aromatic compounds, the bands being various combination modes of fundamental vibrations at lower wavenumbers. In this particular example the exact pattern is not obviously recognizable but, since it is quite unlike the pattern usually observed with mono-substituted aromatic compounds, these are probably ruled out.

The bands at 1620 and 1590 cm^{-1} suggest an aromatic ring, although, particularly in presence of a nitrogen atom, exact assignments are not possible.

Of the three bands in the 1450 cm^{-1} region (1475, 1455 and 1445 cm^{-1}) at least one of them probably arises from the scissoring motion of methylene groups which were detected by the bands in the 2900 cm^{-1} region. It should also be noted that the antisymmetric band of a methyl group would also appear in this region (1460 cm^{-1}) while the symmetric mode would appear at 1378 cm^{-1}. Note the band at 1375 cm^{-1} in the present example. It has already been noted above that no methyl C—H stretching frequencies can be seen (2960 cm^{-1}) and, despite the bands at $\sim$1400 cm^{-1}, methyl groups appear to be contraindicated. For improved resolution in the 3000 cm^{-1} region, however, either a LiF or CaF_2 prism, or better, a grating, instrument is required, while the present spectrum has been run with sodium chloride optics. It would be advisable therefore to

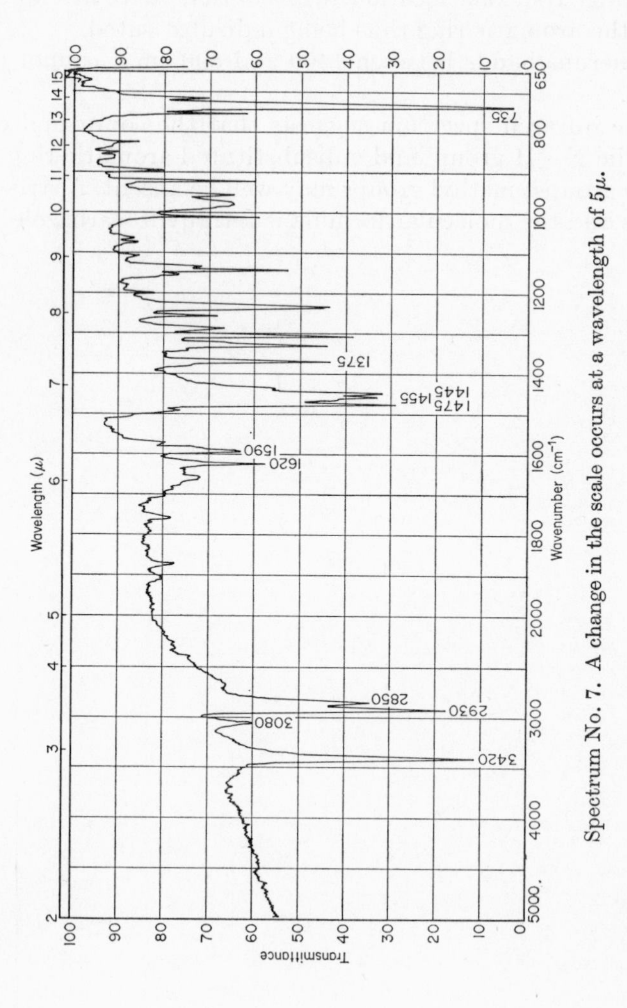

Spectrum No. 7. A change in the scale occurs at a wavelength of 5μ.

re-examine the C—H stretching region under better resolving conditions before concluding finally that methyl groups are completely absent.

The strongest band in the spectrum occurs at 735 cm^{-1} and this arises from a C—H out-of-plane deformation of the H atoms attached to the aromatic ring. Its exact location suggests that there are four adjacent H atoms, the aromatic ring thus being *o*-disubstituted.

The numerous bands between 1400 and 800 cm^{-1} cannot readily be assigned.

Thus the infrared spectrum suggests that the compound contains a rather acidic N—H group, and *o*-disubstituted aromatic ring and some methylene groups; methyl groups may well be absent. A structure to fit these facts and the molecular formula is tetrahydrocarbazole.

Spectrum No. 8

This substance is a yellow crystalline material $C_6H_6N_2O_2$ which absorbs strongly in the near ultraviolet. Sample presentation: potassium bromide disk (0·5 mg/200 mg KBr).

The fact that a yellow compound contains N suggests immediately that the substance is an aromatic nitro compound. Because the presence of O and N considerably modifies the interpretation of an infrared spectrum, the possibility of a nitro compound will be borne in mind during the initial interpretation of the spectrum.

Three prominent bands are observed in the 3000–4000 cm^{-1} region. They are at 3500, 3380 and 3230 cm^{-1}. If we assume that all the oxygen in the molecule is contained in the nitro group, none of these bands can be due to an O—H stretch and all three must arise from an NH or NH_2 group. Since an NH group normally gives rise to one band and an NH_2 group to two (symmetrical and antisymmetrical stretch), it is simplest to assume that the two bands at higher wavenumbers (3500 and 3380 cm^{-1}) arise from an NH_2 group. The third band at 3230 cm^{-1} could arise from some of the molecules being hydrogen bonded so that they would absorb at lower wavenumbers.

A weak band is just apparent at 3100 cm^{-1} indicative of aromatic C—H groups.

The presence of an aromatic ring is confirmed by the pattern of weak bands in the range 1650–2000 cm^{-1}. As this pattern is characterized by a single weak but prominent band at 1930 cm^{-1} 1,4-disubstitution is suggested although this evidence is not very strong.

The bands at 1630, 1590, 1480 and 1450 cm^{-1} are difficult to assign individually in such a complex molecule, but one would expect strong bands in this region arising primarily from the N—H deformation, C=C stretch and the antisymmetrical N—O stretching modes respectively. Interactions may occur so that the exact locations are difficult to anticipate and some of the bands may only be assigned very approximately to a particular mode.

The broad intense band near 1300 cm^{-1} probably arises from the N—O antisymmetrical stretching modes.

The remaining bands at 1175, 1008, 840, 750 and 690 cm^{-1} are of doubtful value for diagnostic purposes although it may be noted that the 840 cm^{-1} band could be due to the aromatic C—H out-of-plane deforma-

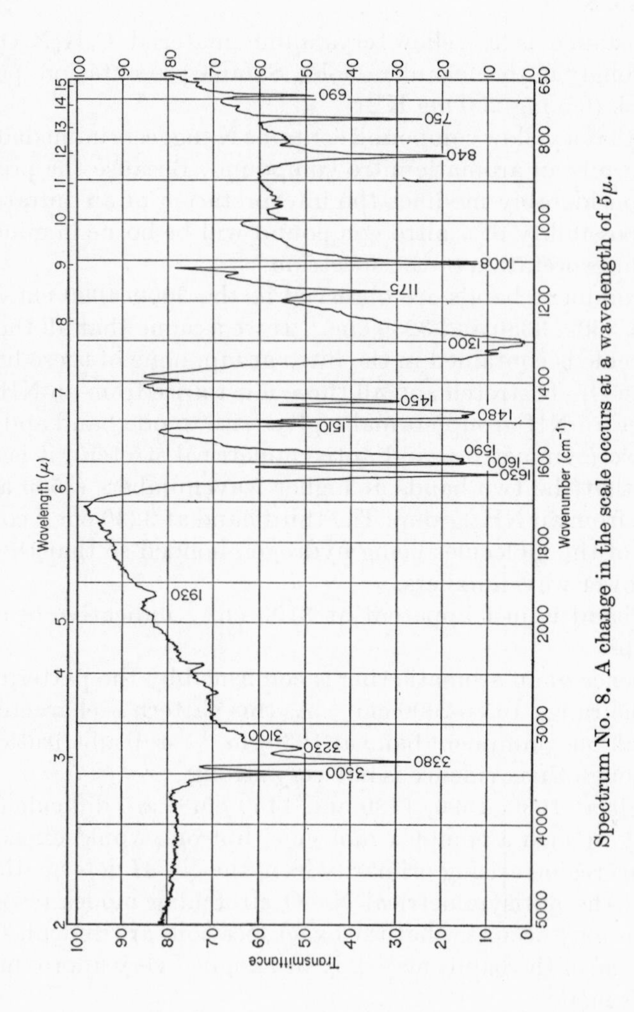

Spectrum No. 8. A change in the scale occurs at a wavelength of 5μ.

tions. Its location suggests two adjacent hydrogen atoms which would strengthen the possibility of 1,4- or *p*-substitution tentatively proposed from the aromatic combination region.

Thus the evidence from the yellow colour, molecular formula and infrared spectrum suggests that the compound is a nitroaniline, probably the *p*-isomer.

Spectrum No. 9

A colourless oil. Sample presentation: capillary film between rock salt plates.

From its immediate appearance, the very strong bands in the region 1400–700 cm^{-1}, suggest that a large number of polar groups are present. It should be remembered that absorption intensity is more properly expressed as the area under the band (cf. spectrum No. 1): this is fundamentally connected to the rate of changes of dipole moment with respect to the changes in distortion of the bond lengths and bond angles in that part of the molecule in which the vibration is largely localized. Vibration in the 1400–700 cm^{-1} region is mainly that of the skeleton of the molecule and where absorption is intense as in the present example, the spectrum is suggestive of highly polar groups such as esters or ethers. In hydrocarbons for example the molecular skeleton has no appreciable dipole in either the distorted or undistorted state and there is therefore but a small change in dipole: bands are consequently weak. Now, ancillary information states that this substance is insoluble in all the usual solvents and this is consistent with a certain inorganic content to the molecule. With the very widespread use of silicon polymers for lubricating ground glass surfaces or for gas chromatography stationary phases, it is the common experience of infrared spectroscopists that silicon greases occasionally contaminate samples submitted by the organic chemist.

The band at ~3100 cm^{-1} may be assigned to aromatic or olefinic C—H stretching vibrations, that at 2950 cm^{-1} to the same vibration with saturated carbon.

The band at 1260 cm^{-1} may be assigned to the Si–CH_2 deformation vibration, that at 800 cm^{-1} to the Si—CH_3 rocking mode. The very intense peaks at 1090–1020 cm^{-1} are due to Si—O—Si and Si—O—C stretching frequencies. Aromatic groups in this material are further suggested by peaks at 1430 and 1130 cm^{-1}, characteristic of the silicon phenyl link, and this is upheld by the normal monosubstituted aromatic C—H out-of-plane deformations, 735 and 700 cm^{-1}. The skeletal in-plane vibrations at 1500 and 1600 cm^{-1} are to be noted though these last are rather low in intensity.

One practical point is worth mention. It has been found that, if potassium bromide is stored in a vacuum desiccator lubricated with silicones, contamination of the salt can occur sufficient to ruin the batch of potassium bromide for disk purposes. The group of very intense bands at 1150–1000 cm^{-1} is responsible for the worst of this effect (cf. B., chap. 20).

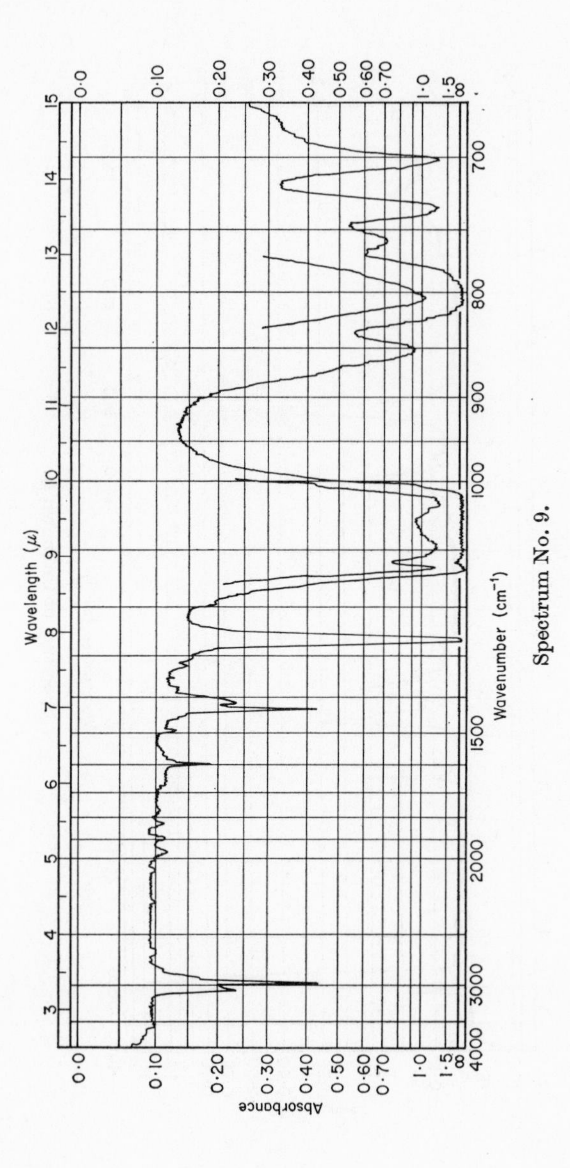

Spectrum No. 9.

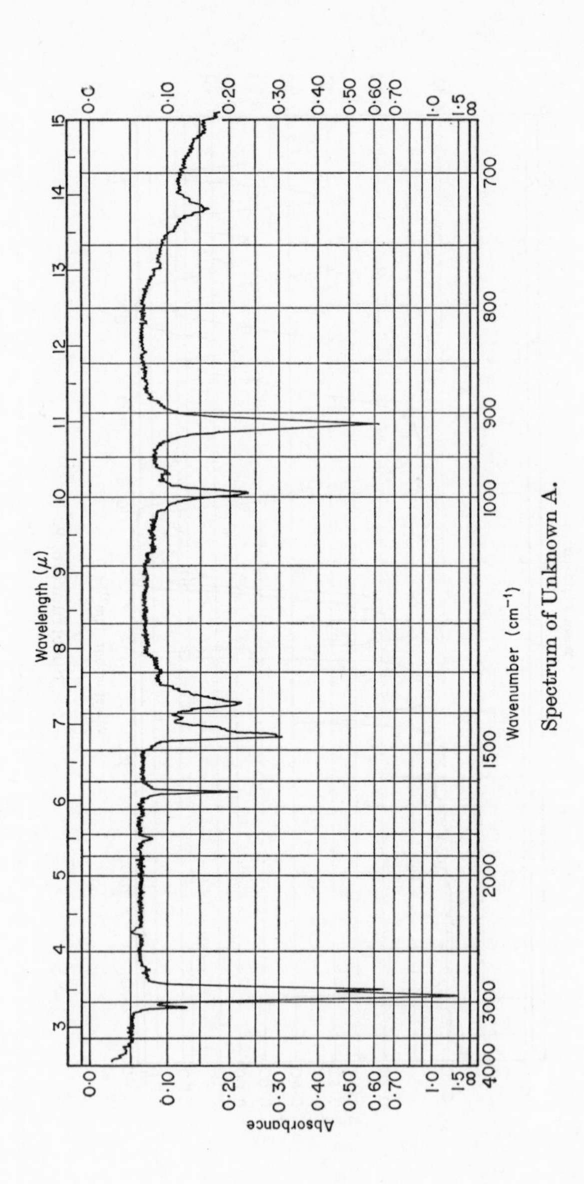

Spectrum of Unknown A.

Unknown A

A liquid hydrocarbon. Sample presentation: capillary film between rock salt plates.

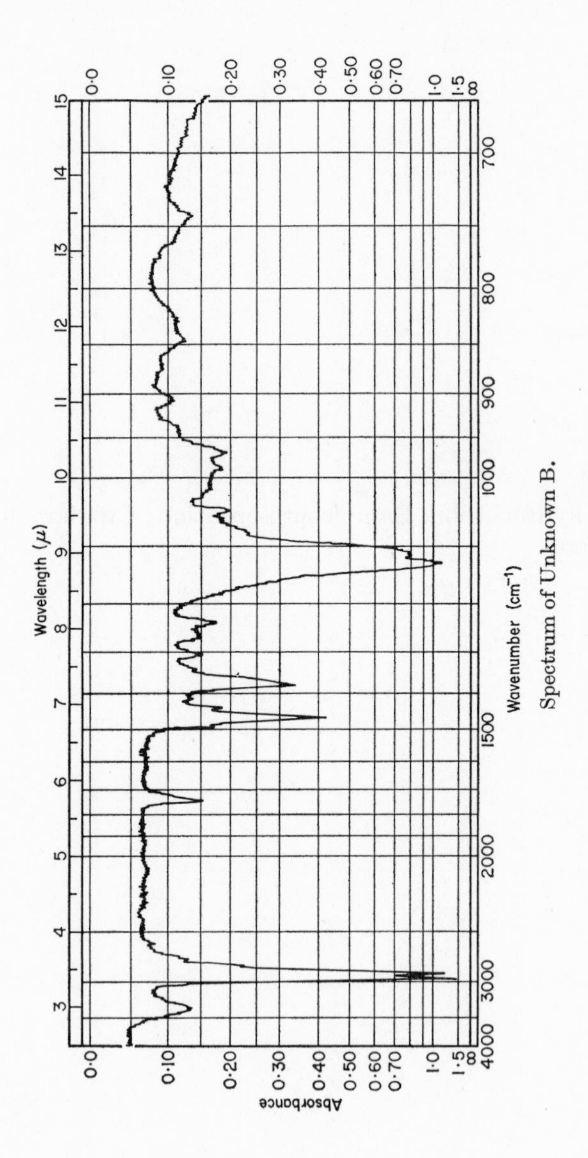

Spectrum of Unknown B.

Unknown B

$C_8H_{18}O$. Sample presentation: capillary film between rock salt plates.

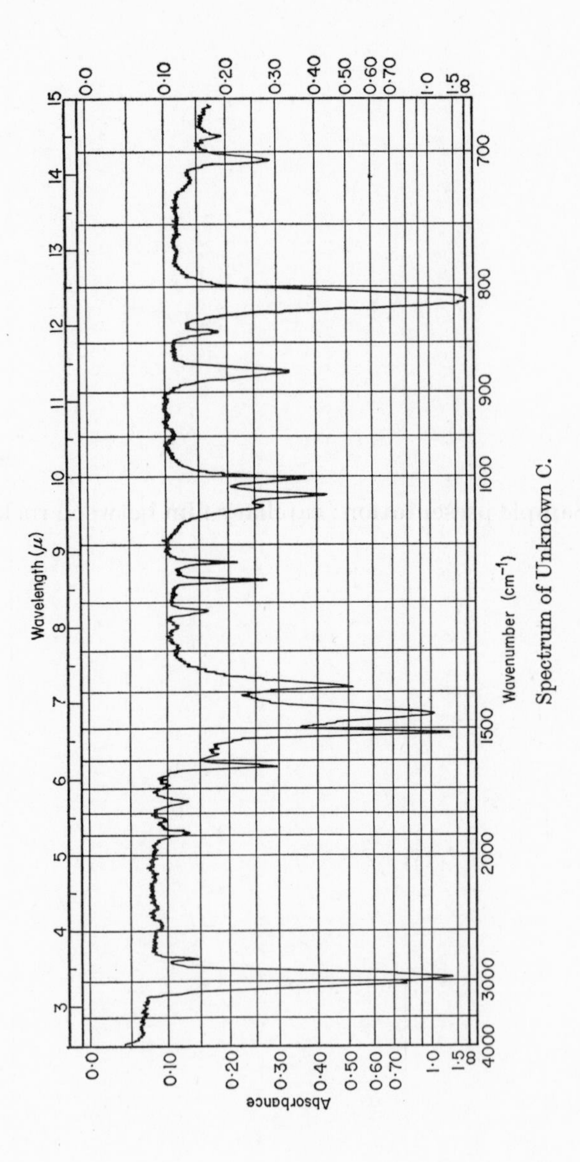

Spectrum of Unknown C.

Unknown C

This substance has a molecular formula C_9H_{12} and absorbs in the near ultraviolet. Sample presentation: pure liquid at a path length of 0·025 mm.

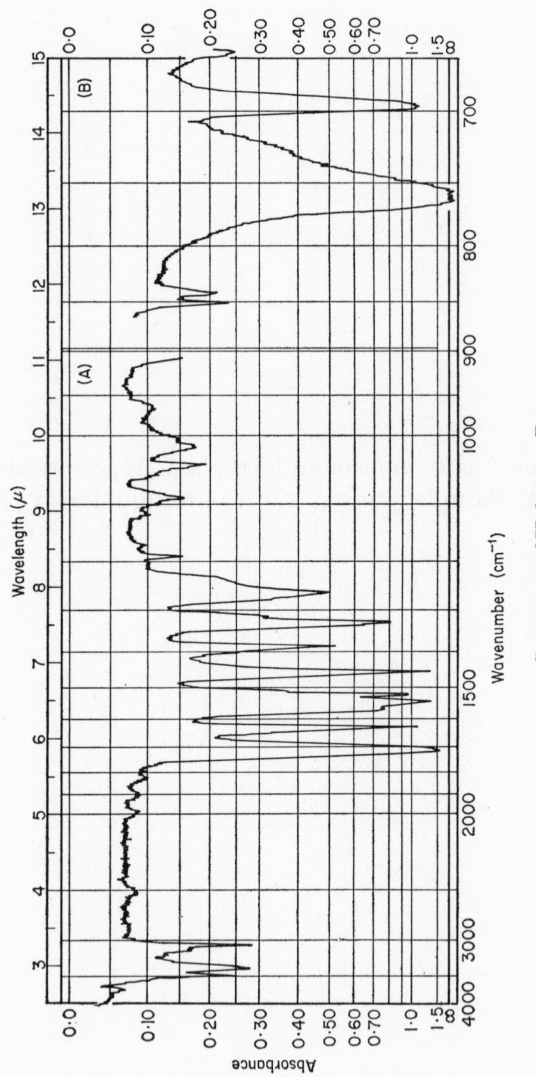

Spectrum of Unknown D.

Unknown D

This substance has a molecular formula C_8H_9ON. Sample presentation: A, 20% solution in chloroform at a path length of 0·1 mm; B, Nujol mull.

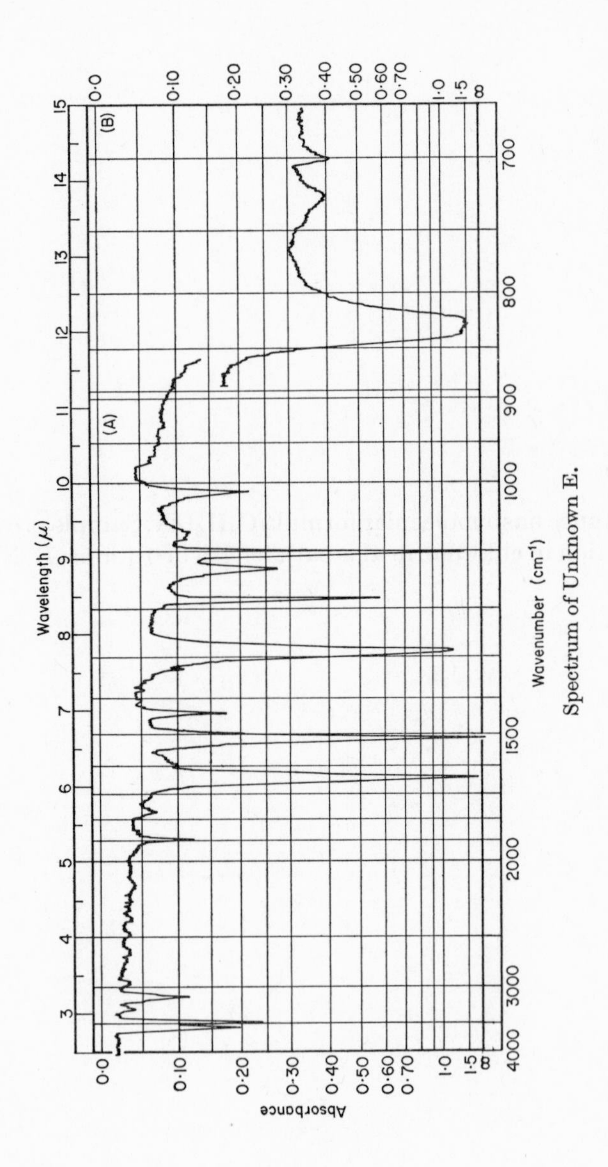

Spectrum of Unknown E.

Unknown E

Molecular formula C_6H_6NCl. Sample preparation: A, carbon tetrachloride at a path length of 0·1 mm; B, Nujol mull.

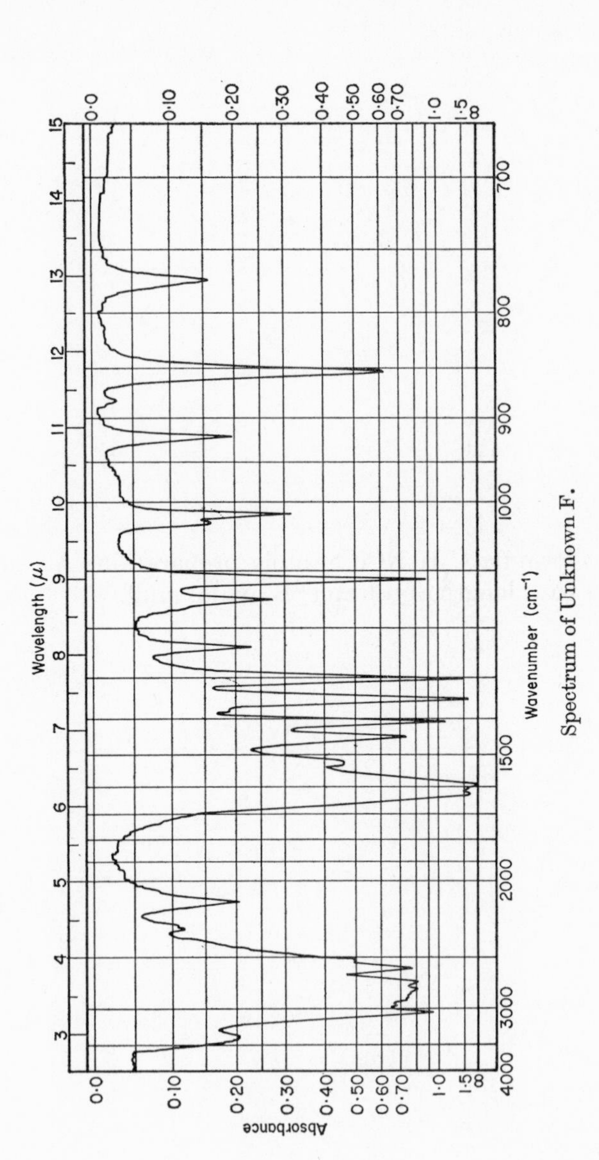

Spectrum of Unknown F.

Unknown F

A high-melting crystalline material soluble in water. It contains C, H, O and N. Sample presentation: potassium bromide disk (0·7 mg/300mg KBr).

Bibliography

The following is a selection of recent books dealing with organic applications of infrared spectroscopy.

Bellamy, L. J. (1948). "Infra-red Spectra of Complex Molecules", 2nd Ed. Wiley, New York.

Jones R. N. and Sandorfy C. (1956). Application of Infrared and Raman spectra *In* "Chemical Applications of Spectroscopy", ed. by W. West. Interscience, New York.

Nakanishi, Koji (1962). "Infra-red Absorption Spectroscopy". Holden-Day, San Francisco.

Silverstein, R. M. and Bassler, G. C. (1963). "Spectrometric Identification of Organic Compounds". Wiley, New York.

Brügel, W. (1962). "An Introduction to Infra-red Spectroscopy", translated by Katritzky. Methuen, London.

Meloan, C. E. (1963). "Elementary Infra-red Spectroscopy". Macmillan, London.

Katritzky, A. R. and Ambler, A. P. (1963). Infrared Spectra. *In* "Physical Methods in Heterocyclic Chemistry", ed. by A. R. Katritzky, Vol. II. Academic Press, New York.

Barrow, G. M. (1963). "The Structure of Molecules: an Introduction to Molecular Spectroscopy". Benjamin, New York.

Cole, A. R. H. (1963). Applications of Infrared Spectroscopy. *In* "Elucidation of Structures by Physical and Chemical Methods", ed. by K. W. Bentley. Part I. Interscience, New York.

Cross, A. D. (1960). "Introduction to Practical Infra-red Spectroscopy". Butterworths, London.

Davies, M. (ed.) (1963). "Infrared Spectroscopy and Molecular Structure". Elsevier, Amsterdam.

Flett, M. S. (1963). "Characteristic Frequencies of Chemical Groups in the Infra-red". Elsevier, Amsterdam.

Potts, W. J. (1963). "Chemical Infra-red Spectroscopy". Wiley, London.

Szymanski (ed.) (1963). "Infra-red band handbook". Plenum Press, New York.

Rao, C. N. R. (1963). "Chemical Applications of Infra-red Spectroscopy". Academic Press, New York.

Mass Spectrometry

Introduction

The demonstration that even a complicated molecule can give rise to a well-defined, reproducible mass spectrum has led to a renewed interest in the application of mass spectrometry to the investigation of organic structures. Whilst the subject is not yet on a systematic footing, there is nevertheless sufficient empirical information about the cracking patterns of organic molecules to enable at least partial structural determinations to be made successfully. The spectra which follow have been discussed from this viewpoint.

Mass spectra are seldom used in isolation, and much other evidence, chemical and physical, is taken into account in their interpretation. Thus, knowledge of the elements present in a compound would have simplified considerably the discussions of some of the spectra which follow. For purely pedagogic reasons, however, little or no ancillary information of this sort has been used.

Some general points must continually be kept in mind during interpretation. First, the only observable properties of compounds upon electron impact are the masses* and abundances of the ions formed.

Secondly, in all the molecules discussed, the proportions of isotopes present are assumed to be those of their natural abundances; in none has artificial enrichment been introduced. It is important to differentiate clearly between the natural abundance and the proportions of the isotopes actually present in the molecule (ratio of abundances of ions). Thus, the percentage abundances of the stable isotopes of carbon ^{12}C and ^{13}C are 98·9 and 1·11, respectively. In the parent molecular ion of ethane, therefore, there should be two isotopic ions at m/e 31 and 32 corresponding to $^{12}C^{13}CH_6$ and $^{13}C_2H_6$. Relative to $^{12}C_2H_6$ expressed as 100%, the abundance of these two ions, calculated by the binomial theorem, is 2·2 : 0·0, respectively (see p. 147). Problems raised by oxygen and chlorine isotopes will be discussed as they arise in the text.

Thirdly, ion–molecule reactions may occur in the ion source, especially if the gas pressure is high, or if the molecule is fairly polar. In the case of the parent molecular ion $[P]^+$, the result of an ion–molecule reaction is an ion one mass unit heavier, and the peak $[P+1]^+$ will of course coincide with that arising from the ^{13}C isotope present. This ion therefore appears

* Strictly the mass to charge ratio (m/e), since the charge is usually unity, however, this is equivalent to the mass itself.

more abundant than would be expected from the isotope contribution alone, and calculations of carbon content based on the abundance of the ion will lead to high values. Accordingly, the number of carbon atoms deduced by this method should be considered as an upper limit only.

Fourthly, some useful considerations allow the detection of an odd number of nitrogen atoms both in the parent molecular ion and in fragment ions. In an organic compound, the commonly occurring atoms, other than carbon and hydrogen, are nitrogen, oxygen, sulphur, fluorine or chlorine and bromine. Nitrogen apart, all these atoms have either odd mass and odd valency, or even mass and even valency; thus, for a molecule with no "free valencies", the molecular weight must be even. Nitrogen is unique amongst the above atoms in having even atomic weight but odd valency. A molecule which has no free valencies but an odd number of nitrogen atoms will have an odd molecular weight; for example, NH_3, $(CH_3)_3N$, $HOCH_2CH_2NH_2$ and $H_6P_3N_3$ which have molecular weights of 17, 59, 61 and 141, respectively. Should the compound possess an even number of nitrogen atoms, then the molecular weight will be even; N_2H_4 has a molecular weight of 32. Equally, a molecule such as NO, which has one nitrogen atom and one free valency is of even mass.

Now fragment ions may be regarded formally as free radicals produced by the dissociation of the original molecule*. Such radicals then lose an electron to give stable even-electron ions. It follows that a univalent radical derived from a molecule and composed of the elements already listed, nitrogen excepted, will have an odd mass: the masses of the radicals $C_2H_5\cdot$, $HCO\cdot$, $\cdot CH_2{}^{35}Cl$ and $\cdot CF_3$ are 29, 29, 49 and 69, respectively. Ionization by removal of an electron makes no observable difference to the weight, and such fragment ions are of odd mass. A similar argument will confirm that fragment ions having an odd number of nitrogen atoms will be of even mass.

Lastly, for guidance, it is useful to reiterate a few general principles. The abundance of the parent molecular ion relative to that of all the ions present is often a useful parameter. It has been found empirically that this abundance is often quite large for carbocyclic ring systems, for aromatic compounds without long side chains and for small straight chain hydrocarbons. The abundance of the ion diminishes (a) with an increase in the number of carbon atoms in the chain, (b) if a centre of branching occurs and (c) if the chain contains a heteroatom. Molecules having a quaternary carbon often do not give a parent molecular ion.

Because of re-arrangements, the ions which occur at the lower-mass end of the spectrum are less useful in deducing structure, although many

* The "molecular ion" is that ion formed from the original molecule by removing one electron only. In the spectrum this gives rise to the "parent peak".

exceptions exist, particularly when the molecule is not large. It is frequently advantageous to concentrate attention firstly on the very abundant ions, beginning with the parent molecular ion and descending the mass scale.

The nomenclature which has been adopted to designate the atoms and bonds involved in the fission processes is as follows:

Atoms	Bonds
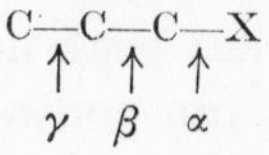 C—C—C—X	C—C—C—X
↑ ↑ ↑	↑ ↑ ↑
γ β α	γ β α

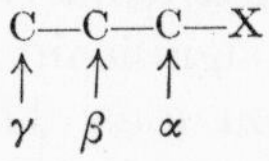

where X represents the functional group.

In the spectra reproduced in the text, thick lines are accorded an intensity ten times that of the thin lines. The strongest peak in the spectrum (base peak) is assigned a value of 100; all other lines are plotted as a percentage of this.

'z-Number' Classification

A valuable concept that helps to codify mass spectra is that of the z-number. For example all hydrocarbons can be given the general formula C_nH_{2n+z}, particular examples being the paraffins, C_nH_{2n+2}, where $z = +2$, and the alkyl benzenes, C_nH_{2n-6}, where $z = -6$. For hydrocarbons, moreover, a mass-number table can be constructed such that all mass numbers corresponding to a given z-number fall in one column and all mass numbers corresponding to a given carbon number fall in one row.

TABLE 1

z-Number table (C_nH_{2n+z})

Carbon No.	z-Number −11	−10	−9	−8	−7	−6	−5	−4	−3	−2	−1	0	+1	+2
1										12	13	14	15	16
2	17	18	19					24	25	26	27	28	29	30
3	31	32	33	34	35	36	37	38	39	40	41	42	43	44
4	45	46	47	48	49	50	51	52	53	54	55	56	57	58
5	59	60	61	62	63	64	65	66	67	68	69	70	71	72
6	73	74	75	76	77	78	79	80	81	82	83	84	85	86
7	87	88	89	90	91	92	93	94	95	96	97	98	99	100
8	101	102	103	104	105	106	107	108	109	110	111	112	113	114
9	115	116	117	118	119	120	121	122	123	124	125	126	127	128
10	129	130	131	132	133	134	135	136	137	138	139	140	141	142

This is possible as the difference between successive homologues is the methylene group, CH_2. Since this corresponds to 14 mass units, an increase of 14 masses will be equivalent to increasing an alkyl chain by one CH_2 group, i.e. one carbon number, without changing the molecular type, described by the z-number. Thus, each z-number column must consist of figures increasing in 14 mass number intervals. Table 1 illustrates the application of these ideas.

The class of a hydrocarbon can be determined from the column in which its parent peak lies, or, where there is no significant parent peak, from the column containing the major fragment ions. This is usually one z-number less than that of the parent molecule ion. The major classes of hydrocarbons coded by z-number are given in Table 2.

TABLE 2

Hydrocarbons by z-number

C_nH_{2n+2}	Paraffins
C_nH_{2n}	Mono-olefins and cyclo-paraffins
C_nH_{2n-2}	Dienes, cyclenes, dicyclanes, acetylenes
C_nH_{2n-4}	Trienes, cyclodienes, etc. etc. tricyclanes
C_nH_{2n-6}	Alkylbenzenes; tetra-cyclanes (steranes)
C_nH_{2n-8}	Indanes and tetralins
C_nH_{2n-10}	Indenes; dicycloalkylbenzenes
C_nH_{2n-12}	Naphthalenes (overlaps C_nH_{2n+2})
C_nH_{2n-14}	Biphenyls (overlaps C_nH_{2n})
C_nH_{2n-18}	Phenanthrenes/Anthracenes (overlaps C_nH_{2n-4})

This concept can also be extended to substances other than hydrocarbons and a brief outline is given below.

In the case of amines, for example, as we pass from propane to propylamine ($C_3H_8 \rightarrow C_3H_7NH_2$) this represents the replacement of a H atom by NH_2. Nitrogen has the same *mass* as CH_2 so that, in terms of equivalent mass, we now have a hydrocarbon $C_3H_7 + CH_2 + H_2$, i.e. C_4H_{11}, of general formula C_nH_{2n+3}. In equating the original $C_3H_7NH_2$ to C_4H_{11} we have increased the carbon number by one, and the general formula may now be written $C_{n+1}H_{2(n+1)-z}$, equivalent in mass to C_nH_{2n+3}.

In terms of mass units

$$C_nH_{2n+3} = 12n+2n+3 = 14n+3$$

$$C_{n+1}H_{2(n+1)-z} = 12n+12+2n+2-z = 14n+14-z$$

$$\therefore z = 11$$

By the same kind of argument, alcohols have a z-number of -10. Table 3 shows the z-numbers for a few substances other than hydrocarbons. The classification begs the question of deducing carbon number from the parent peak in the case of compounds other than hydrocarbons.

TABLE 3

Non-hydrocarbons by z-number

Equivalent z-number	
-10	Alcohols, acids, esters, ethers
$+2$	Ketones, aldehydes
-11	Amines; quinolines; amides
-5	Anilines; pyridines
-3	Pyrroles
-8	Alkyl sulphides; mercaptans
-6	Benzthiophenes
0	Thiophenes
-4	Phenols
-2	Thiophenols

If x is the apparent carbon number obtained from Table 1, we must subtract the carbon equivalent of each heteroatom in the compound ($O = 1$, $N = 1$, $S = 2$) and the carbon number is then one less than the resulting figure. Thus, for $C_3H_7NH_2$ (MW 59), the apparent carbon number from Table 1 is 5: subtract 1 for the carbon equivalent of N and the true carbon number is then 3.

Use of this "z-number system" will be made in the discussion of the first five spectra.

Mass Spectra and their Interpretation

Spectrum No. 1

The apparent parent peak P (and hence molecular weight) is at *m/e* 114. The ratio of the intensities of the peaks at *m/e* 115 and 114 is 0·32/3·82 = 8·4%. This is the correct isotope ratio of $^{13}C/^{12}C$ for 8 carbon atoms per molecule. Further, since the parent peak falls in the $z = +2$ series, the compound is probably a paraffin and, hence, an octane.

Other possible compounds include a heptanone or a heptaldehyde, and, in addition, cyclic or unsaturated, oxygen-containing compounds of empirical formula $C_7H_{14}O$. These may be excluded on the grounds of isotope ratios (^{17}O natural abundance is negligible hence the ratio of 115/114 would be about 7%). Ketones[1] and aldehydes[2] tend to re-arrange to give fairly intense peaks on the $z = +2$ series at lower masses by the following fragmentation:

$$CH_2CH_2 \,|\, CH_2C(=O)CH_3 \;(\text{H transferred to O}) \longrightarrow CH_2{=}CH_2 + [CH_2C(OH)CH_3]^+$$

However, no intense peaks are observed at *m/e* 30, 44, 58 ... 100 on this series. Oxygenated compounds in general, moreover, are excluded by the absence of significant peaks at *m/e* 31 and 45.

Confirmation that the compound is a paraffin is provided by the intense peaks at *m/e* 43, 57, 71 and 99 corresponding to the C_nH_{2n+1} series for C_3 to C_7. Accepting that the compound is an octane, we may try to identify which of the 18 isomers is present. The octanes have been discussed in detail by Beynon[3a] and by Bloom, Mohler and their co-workers.[4]

The absence of an intense peak *m/e* 85 $[P-29]^+$ implies the inability to lose an ethyl group, and hence the absence of an ethyl group in the molecule. The only compounds to satisfy this condition are $25M_2C_6$,* $234M_3C_5$, $224M_3C_5$ and $2233M_4C_4$. Of these $224M_3C_5$ and $2233M_4C_4$ have t-butyl groups which would readily dissociate from the molecule, and, appearing at *m/e* 57, would constitute the base peak. Although the peak at *m/e* 57 is certainly an intense one (~80%), it is not the base peak and

* The shortened nomenclature $25M_2C_6$ has been used to indicate 2,5-dimethylhexane (see Table 4), etc.

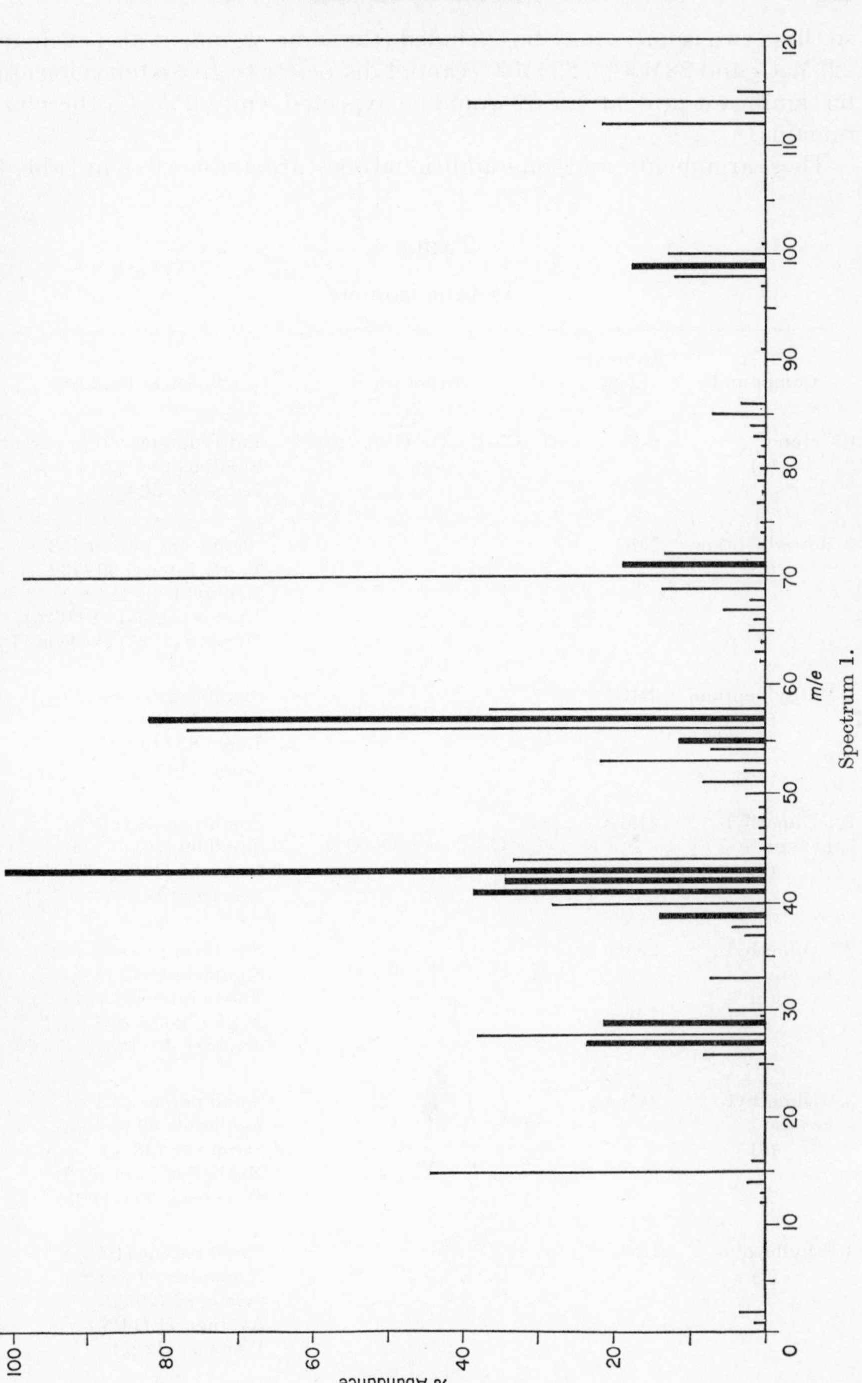
% Abundance
100
80
60
40
20
0
m/e
10
20
30
40
50
60
70
80
90
100
110
120

Spectrum 1.

so these two isomers may be excluded (the same argument also excludes $223M_3C_5$ and $22M_2C_6$). $234M_3C_5$ cannot dissociate to give a butyl ion and no significant peak at *m/e* 57 would be expected. Only $25M_2C_6$ therefore remains.

These arguments, and some additional ones, are summarized in Table 4.

TABLE 4

Octane isomers

Compound	Abbreviation	Structure	Spectral Features
n-Octane (1)	n-C_8	C—C—C—C—C—C—C—C	Fairly intense (7%) parent Small 99 (0·1%) Large 85 (30%)
2-Methylheptane (2)	$2MC_7$	C—C(—C)—C—C—C—C—C	Significant parent (5%) Fairly intense 99 (13%) Negligible 85 (1·8%) Unexceptional 71 (13%) 70 peak (17%) exceeds 71
3-Methylheptane (3)	3MC7	C—C—C(—C)—C—C—C—C	Significant parent (3%) Small 99 (0·8%) Large 85 (49%) Small 71 (3%)
2,4-Dimethyl-hexane (4)	$24M_2C_6$	C—C(—C)—C—C(—C)—C—C	Small parent (1·7%) Small 99 (1%) Large 85 (46%) Unexceptional 71 (15%)
2,5-Dimethyl-hexane (5)	$25M_2C_6$	C—C(—C)—C—C—C(—C)—C	Significant parent (4%) Significant 112 (2%) Fairly intense 99 (17%) Negligible 85 (0·7%) Average 71 (19%)
3,4-Dimethyl-hexane (6)	$34M_2C_6$	C—C—C(—C)—C(—C)—C—C	Small parent (2·2%) Negligible 99 (0·3%) Strong 85 (38%) Negligible 71 (1·5%) Base peak 56 (100%)
3-Ethylhexane (7)	$3EC_6$	C—C—C(—C—C)—C—C—C	Small parent (1·6%) Negligible 99 (0·1%) Strong 85 (29%) Average 71 (14%) Weak 57 (12%)

TABLE 4—*continued*

Compound	Abbreviation	Structure	Spectral Features

2-Methyl-3-ethyl-pentane (8)	$2M3EC_5$	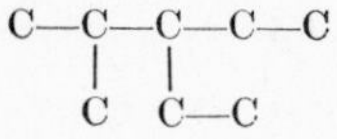 C—C—C—C—C
		   | |
		   C C—C
			Small parent (1·3%)
			Negligible 99 (0·1%)
			Average 85 (18%)
			Strong 71 (25%)
			Even stronger 70 (50%)
			Weak 57 (15%)

3-Methyl-3-ethyl-pentane (9)	$3M3EC_5$	C
		      |
		C—C—C—C—C
		      |
		      C—C
			No parent (0%)
			Small 99 (2%)
			Large 85 (64%)
			Negligible 71 (0·5%)
			Only medium 57 (27%)

2,2-Dimethyl-hexane (10)	$22M_2C_6$	C
		   |
		C—C—C—C—C—C
		   |
		   C
			Negligible parent (0·03%)
			Significant 99 (6%)
			Negligible 85 (0·02%)
			Negligible 71 (0·7%)
			Base 57 (100%)
			Small 43 (16%)

2,2,3-Trimethyl-pentane (11)	$223M_3C_5$	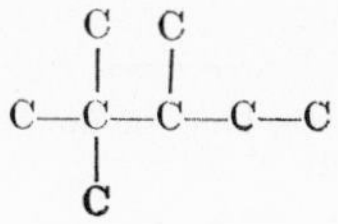 C C
		   | |
		C—C—C—C—C
		   |
		   C
			Negligible parent (0·03%)
			Small 99 (3%)
			Small 85 (3%)
			Negligible 71 (0·4%)
			Base 57 (100%)
			Weak 43 (23%)

2,2,4-Trimethyl-pentane (12)	$224M_3C_5$	C
		   |
		C—C—C—C—C
		   | |
		   C C
			Negligible parent (0·02%)
			Significant 99 (5%)
			Negligible 85 (0·01%)
			Negligible 71 (0·8%)
			Base 57 (100%)
			Weak 43 (23%)

2,2,3,3-Tetra-methylbutane (13)	$2233M_4C_4$	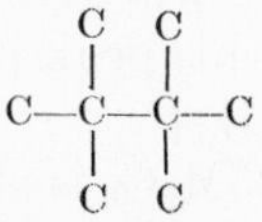 C C
		   | |
		C—C—C—C
		   | |
		   C C
			Negligible parent (0·03%)
			Significant 99 (6%)
			Negligible 85 (0·03%)
			Negligible 71 (0·3%)
			Base 57 (100%)
			Small 43 (18%)

3,3-Dimethyl-hexane (14)	$33M_2C_6$	C
		      |
		C—C—C—C—C—C
		      |
		      C
			Negligible parent (0·01%)
			Significant 99 (5%)
			Strong 85 (36%)
			Strong 71 (47%)
			Medium 57 (41%)

2,3,3-Trimethyl-pentane (15)	$233M_3C_5$	C
		      |
		C—C—C—C—C
		   | |
		   C C
			Negligible parent (0·01%)
			Significant 99 (4%)
			Fairly strong 85 (25%)
			Strong 71 (45%)
			Medium 70 (36%)
			Medium 57 (36%)

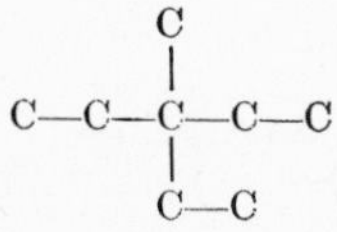

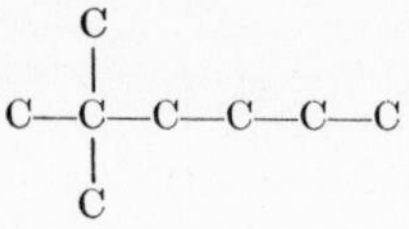

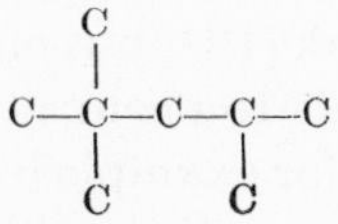

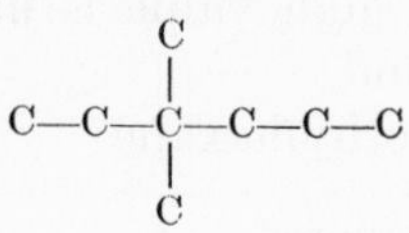

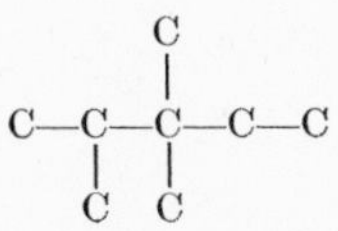

TABLE 4—*continued*

<table>
<tr><th>Compound</th><th>Abbreviation</th><th>Structure</th><th>Spectral Features</th></tr>
<tr><td>4-Methylheptane
(16)</td><td>$4MC_7$</td><td>C
|
C—C—C—C—C—C—C</td><td>Significant parent (3%)
Small 99 (1%)
Small 85 (5%)
Large 71 (53%)
Strong 70 (46%)
Weak 57 (14%)</td></tr>
<tr><td>2,3-Dimethyl-
hexane
(17)</td><td>$23M_2C_6$</td><td>C
|
C—C—C—C—C—C
|
C</td><td>Significant parent (2%)
Negligible 99 (0·4%)
Small 85 (2%)
Strong 71 (46%)
Stronger 70 (58%)
Weak 57 (17%)</td></tr>
<tr><td>2,3,4-Trimethyl-
pentane
(18)</td><td>$234M_3C_5$</td><td>C C C
| | |
C—C—C—C—C</td><td>Weak parent (0·3%)
Negligible 99 (0·2%)
Negligible 85 (0·2%)
Strong 71 (62%)
Strong 70 (41%)
Weak 57 (16%)</td></tr>
</table>

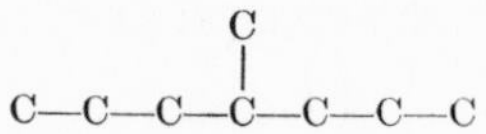

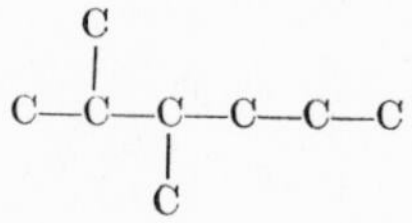

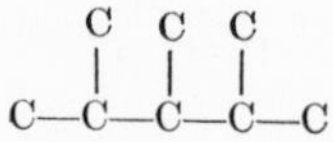

N.B. Unless otherwise stated, *m/e* 43 is the base peak (100%). Note similarity of spectra (10)–(13), of (14) and (15) and of (16)–(18).

Other points which may be noted from Table 4 are: (i) the normal isomer has the biggest parent peak $[P]^+$, but only a weak $[P-15]^+$ due to loss of CH_3. Fragmentation tends to occur at a branched carbon atom and domination of the spectrum for example by a t-butyl group at *m/e* 57 has already been referred to above. Only those octane isomers with a terminal isopropyl group, moreover, will give rise to a significant peak at *m/e* 71, $[P-C_3H_7)^+$. Not only is this peak present in spectrum No. 1, but the base peak lies at *m/e* 43, $[C_3H_7]^+$.

Confirmation of the structure as $25M_2C_6$ is obtained by the exceptionally large peak at *m/e* 112.* The ratio of the ion intensities at masses 112 and 114 is about 53%. This high value is unique for $25M_2C_6$ and is characteristic of that compound.

The sample is thus 2,5-dimethylhexane.

* $[P-2]^+$: exact structure not determined.

Spectrum No. 2

The parent peak at m/e 106 is on the C_nH_{2n-6} series and implies an alkyl benzene: the intensity of this peak is characteristic of aromatic compounds. For confirmation the base peak at m/e 77 suggests a phenyl group, as do the peaks at m/e 39, 50, 51 and 52.

The 39 and 52 peaks probably correspond to $[CH{=}CHCH{=}]^+$ and $[CH{=}CHCH{=}CH]^+$ from simple fragmentation of the benzene ring. Such structures will be strongly resonance stabilized.

Alkyl benzenes of molecular weight 106 include the isomeric xylenes and ethyl benzene. However, all these have a very intense peak at m/e 91 probably due to the tropylium ion;[5] such a peak is absent from spectrum No. 2. Hence these four alkyl benzenes are excluded and we must look for something of similar mass among the non-hydrocarbons.

The molecule readily loses 29 mass units to give what is apparently the $[C_6H_5]^+$ ion at m/e 77. These 29 mass units must come from something other than a hydrocarbon residue and probably correspond to CHO.

Thus the molecule is evidently benzaldehyde. This is confirmed by the very intense $[P-1]^+$ peak (m/e 105), a characteristic of aldehyde spectra and proably due to loss of the aldehydic hydrogen. The $[P+1]^+$ peak is 7·6% of the parent, indicating only seven carbon atoms per molecule (being an oxygenated compound, the $[P+1]^+$ may be slightly enhanced by an ion–molecule reaction). The mass spectra of aldehydes are discussed by Beynon,[3b] by Gilpin and McLafferty[2] and by Aczel and Lumpkin.[6]

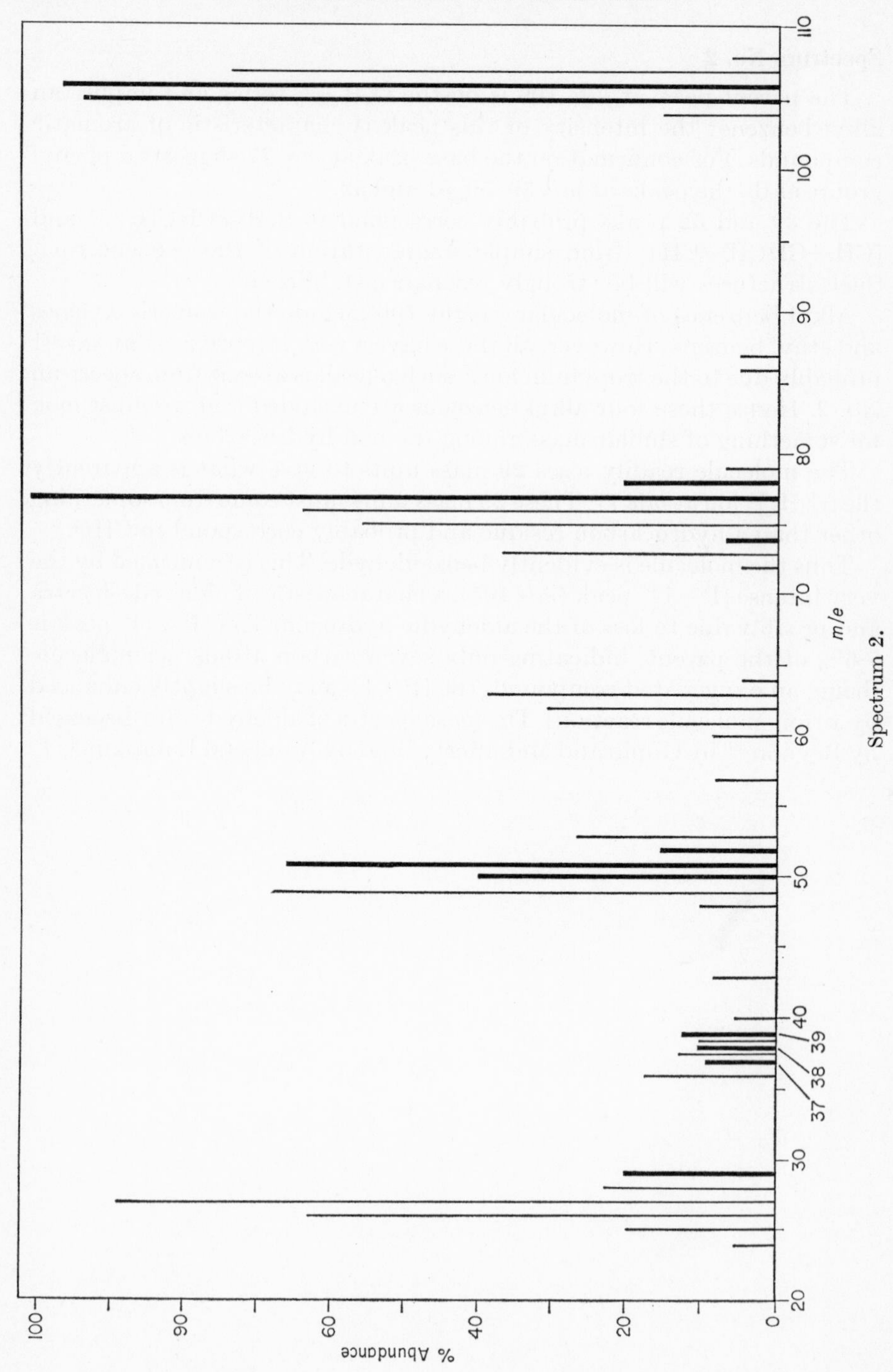
% Abundance
0
20
40
60
80
90
100
m/e
20
30
37
38
39
40
50
60
70
80
90
100
110

Spectrum 2.

Spectrum No. 3

There is evidently a very small parent peak at m/e 88, on the C_nH_{2n-10} series. This suggests an oxygenated compound, borne out by peaks at m/e 73, 45 and 31.

Oxygenated compounds of this molecular weight may be listed thus: eight isomeric pentanols; two isomeric butyric acids; methyl propionate and ethyl acetate; and six isomeric ethers.

The alcohols have been discussed by Friedel and his co-workers.[7] Primary alcohols give hydrocarbon-type spectra, either by dehydration or fragmentation at the bond β to the OH group, where this is also a site of branching. Thus the base peaks of pentan-1-ol, 3-methylbutan-1-ol, 2-methyl-butan-1-ol and 2,2-dimethylpropanol fall on the hydrocarbon series at m/e 42, 41, 57 (or 29)* and 57 respectively.

Secondary and tertiary alcohols split at the point of branching nearest the hydroxyl group, which then carries the positive charge. Thus, with pentan-2-ol and 3-methylbutan-2-ol

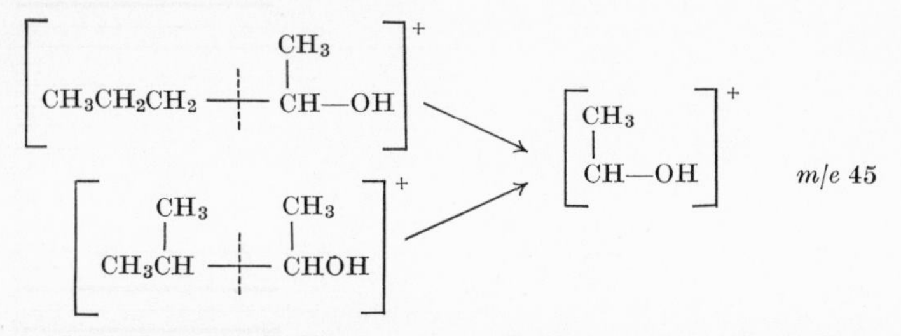

pentan-3-ol and 2-methylbutan-2-ol likewise have their base peaks at m/e 59. Since m/e 73 is the base peak in Spectrum 3, the pentanols are excluded from consideration.

Carboxylic acids were first discussed by Happ and Stewart.[8] Two factors control the fragmentation of the acids: (i) cleavage β to the carboxyl group, with hydrogen re-arrangement; and (ii) cleavage adjacent to the secondary carbon atom, α to the carboxyl group.

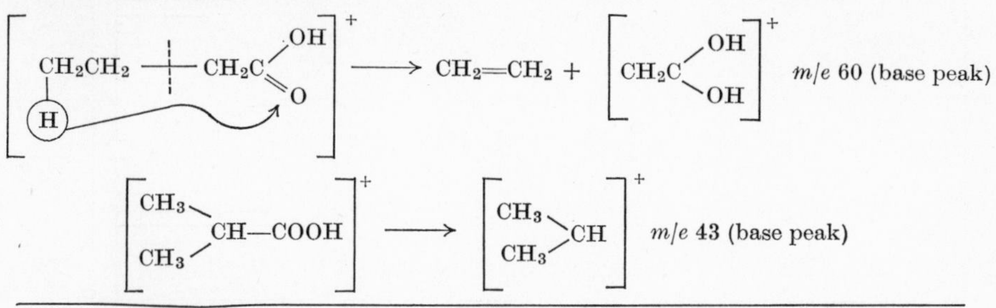

* See footnote p. 139.

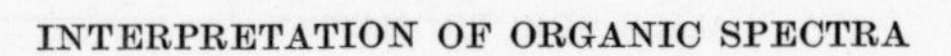

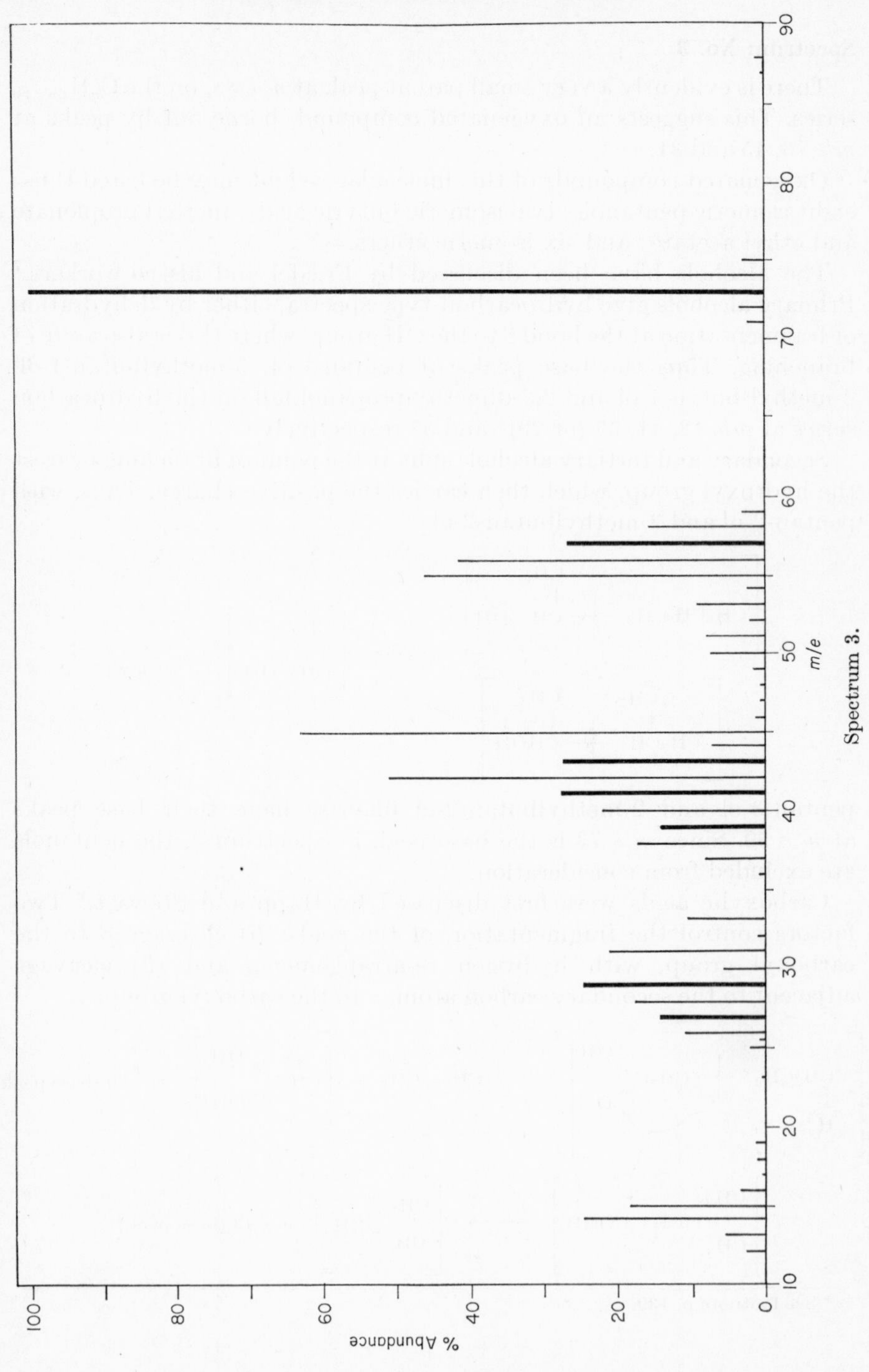

Spectrum 3.

Since the base peak of spectrum 3 is at *m/e* 73, neither n-nor isobutyric acid qualifies for consideration: these are of course the only two acids which meet the requirement for a molecular weight of 88.

The mass spectra of esters have been discussed by Quayle[9] and by Sharkey and his co-workers.[10] The higher esters undergo the following fragmentations.

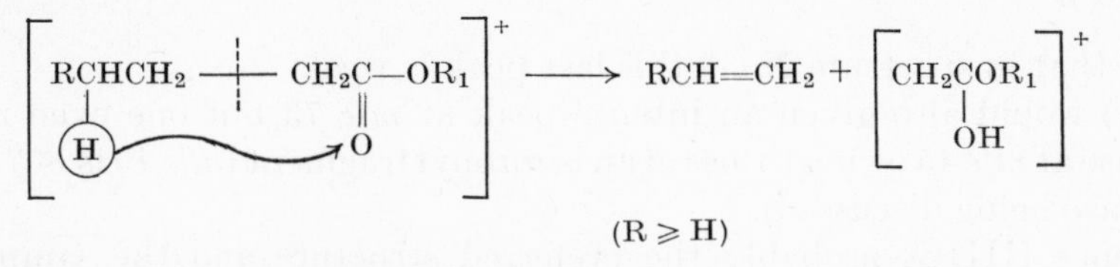

Ethyl acetate and methyl propionate on the other hand undergo cleavage between the carbonyl group and the ester oxygen.

$$[CH_3CO \mid OCH_2CH_3]^+ \longrightarrow [CH_3CO]^+ \quad m/e\ 43\ \text{(base peak)}$$

and

$$[CH_3CH_2CO \mid OCH_3]^+ \longrightarrow [CH_3CH_2CO]^+ \quad m/e\ 57\ \text{(base peak)}$$

Hence neither of these esters fits Spectrum 3 and so, by elimination, we must consider the ethers.

The mass spectra of aliphatic ethers have been discussed by McLafferty.[11]

The base peak at *m/e* 73 $[P-15]^+$ indicates the ready loss of a methyl radical. Since cleavage of the bond β rather than α to the oxygen atom is preferred, two structures may be suggested.

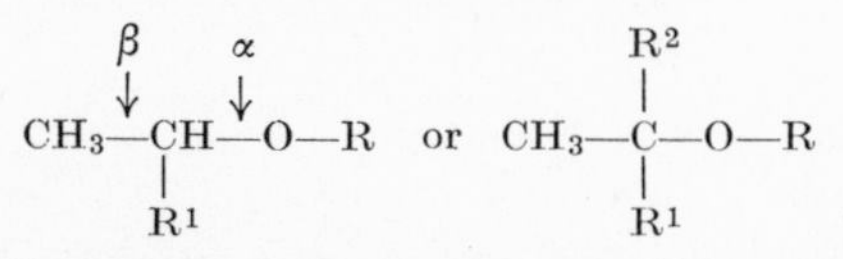

Where R^1 or R^2 is heavier than methyl, it is the larger group which would dissociate rather than the CH_3. Hence for a molecular weight requirement of 88 we probably have

(I) $CH_3CH_2OCH_2CH_2CH_3$ or $CH_3CH_2OCH(CH_3)_2$ (i.e. II below)
(II) $(CH_3)_2CHOCH_2CH_3$
(III) $(CH_3)_3COCH_3$

* In a 180° instrument the 29 peak (CH_3CH_2—) exceeds the 57, but the reverse is true in a 90° mass spectrometer.

(I) would fragment in two ways due to β cleavage, giving intense peaks as shown

$$[CH_3CH_2OCH_2CH_2CH_3]^+ \nearrow [CH_2OCH_2CH_2CH_3]^+ \quad m/e\ 73$$
$$[CH_3CH_2OCH_2CH_2CH_3]^+ \searrow [CH_3CH_2OCH_2]^+ \quad m/e\ 59$$

Note that in spectrum No. 3 this last peak is weak.

(II) would also given an intense peak at m/e 73 but one even more intense at m/e 45 owing to loss of an isopropyl fragment (m/e 45 is $\sim 7\%$ in the case being discussed).

Hence (III) is probably the preferred structure and the sample is methyl t-butyl ether.

Spectrum No. 4

The parent peak at m/e 62 is on the C_nH_{2n-8} series. Since the molecular weight is too low for an indane or a tetralin, the compound must be other than a hydrocarbon. It will therefore contain one sulphur atom per molecule (or two oxygen atoms). The intense (base) peak at m/e 31 implies an oxygenated compound, containing the primary alcohol group CH_2OH. Subtraction of 31 from 62 leaves 31, again, implying a second CH_2OH group, making the molecule $HOCH_2—CH_2OH$ (ethylene glycol). This fulfils the requirement of two oxygen atoms per molecule.

The intense peak at m/e 18 suggests either that the sample was wet or that it dehydrated readily:

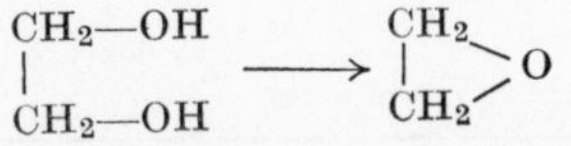

$$\begin{array}{l} CH_2—OH \\ | \\ CH_2—OH \end{array} \longrightarrow \begin{array}{l} CH_2 \\ | \quad \rangle O \\ CH_2 \end{array} \qquad \text{(mass 44)}$$

This is possible, although the peak at m/e 44 is not very significant, being exceeded by those at m/e 45, $[CH_2CH_2OH]^+$ i.e. $[P-OH]^+$ and m/e 43 $[P-H_3O]^+$.

The intense 33 peak is a double re-arrangement, due to protonation of the oxygen atom (comparable to the 61, 75, 89 ... in esters):

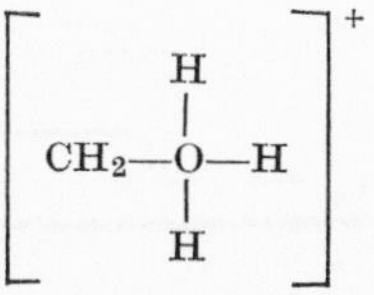

$$\left[\begin{array}{c} H \\ | \\ CH_2—O—H \\ | \\ H \end{array}\right]^+$$

the lesser 32 peak being a single re-arrangement

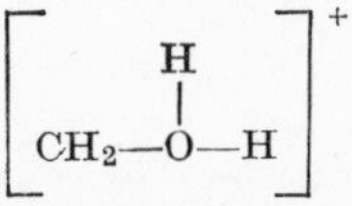

$$\left[\begin{array}{c} H \\ | \\ CH_2—O—H \end{array}\right]^+$$

The compound is ethylene glycol. The mass spectra of diols and alkoxy alcohols have been discussed by Peard and McLafferty.[12]

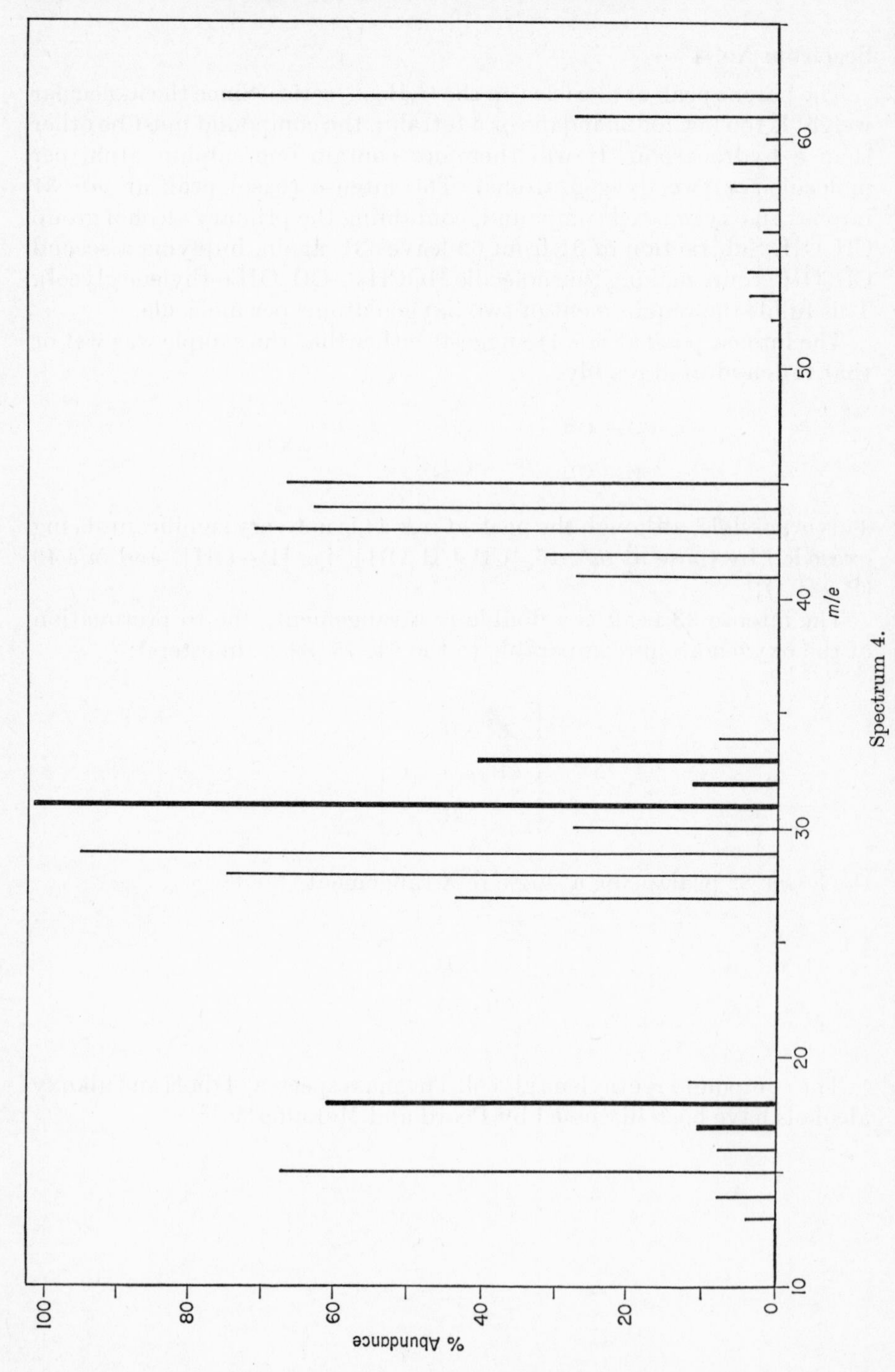

Spectrum 4.

Spectrum No. 5

Preliminary information on this material gives a boiling range of 110–120° C/760 mm. It is stated to contain 3–4% of an impurity.

There is a parent peak (even mass number) at *m/e* 112 (on the C_nH_{2n} series) and, with fragments on the C_nH_{2n-1} series (at *m/e* 97, 83, 69, 55, 41, 27), implies an olefin or a cyclane. A second, weaker parent at *m/e* 92 is to be noted. This lies on the C_nH_{2n-6} series and is therefore an alkyl benzene of which the only representative with this molecular weight is toluene. This is confirmed by the characteristic peak at 91 which is, in fact, the base peak of toluene itself. Toluene is also confirmed by the 91/92 ratio and by the peaks at *m/e* 77, 78, 79. Some peaks at non-integral values are to be noted and these arise from doubly charged ions. Such ions are formed by removal of two electrons by the initial electron impact and, since the charge *e* then equals 2, the ion will be collected and recorded at the mass to charge ratio corresponding to half the absolute mass of the ion. In spectrum 5, the peak at half mass 45·5 due to the doubly charged tolyl ion $[91]^{++}$ should be noted (the 46·5, 47·5, 48·5 and 49·5 must be due to doubly charged ions at *m/e* 93, 95, 97 and 99 from the olefin/cyclane).

Identification of the olefin/cyclane is a much more difficult problem, and may even be impossible by mass spectrometry alone. Although the A.P.I. collection of reference mass spectra[13] lists 19 cyclopentanes and cyclohexanes of molecular weight 112, it includes only 13 of the possible 92 octene isomers.

The parent peak intensity of the major compound in Spectrum 5 is about 16%. The octenes reported range from 7 to 29%, and the cyclanes 1 to 36%. The base peak in the present example is at *m/e* 55 and this is the case for eight of the cylanes and three of the reported octenes. There is only one cyclane with a parent peak between 10 and 20% and the base peak at *m/e* 55 namely *cis*-1-methyl-2-ethylcyclopentane. However, in this case, the 70 peak (53%) exceeds the 69 peak (34%), whereas in Spectrum 5 the 69 peak is roughly twice the 79 peak. It seems reasonable, therefore, to exclude the C_8 cyclopentanes and cyclohexanes, though this could be done more confidently if other evidence were available, e.g. the source of the sample or infrared spectroscopic evidence. Cyclobutanes and cyclopropanes are not considered in the present example and the discussion is continued therefore on the basis that the sample is an octene.

Examining the typical ions formed by fragmentation of an alkyl chain, it is seen that the molecule can lose a methyl group, to *m/e* 97, and an ethyl group to *m/e* 83, thus we have $(C_6H_{11})CH_2CH_3$. Note that the molecule loses a propyl group less readily (*m/e* 69 is less intense than *m/e*

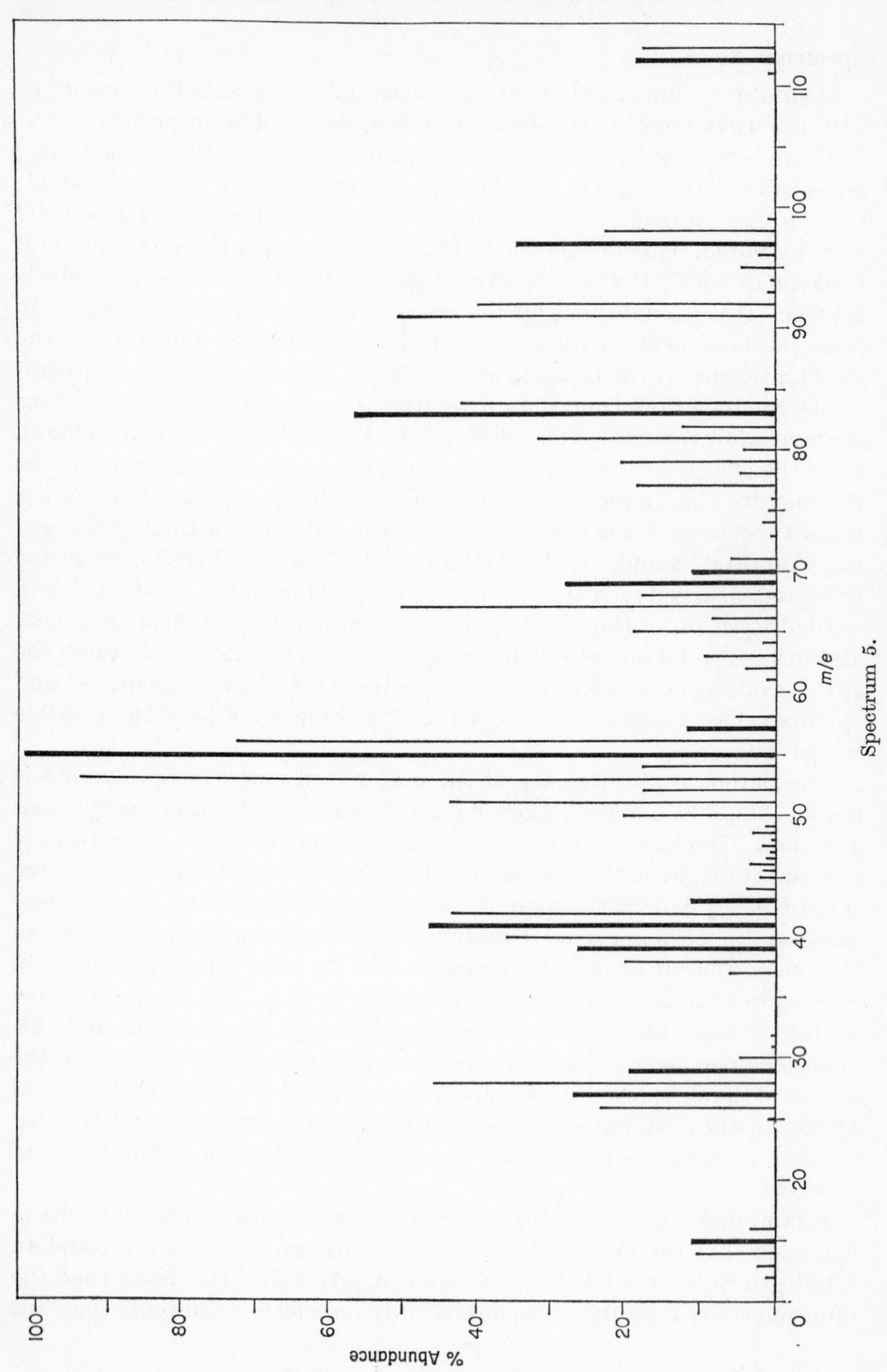

Spectrum 5.

83), so there is probably either a point of branching or a double bond on the next carbon atom. In short, either structure (I) or structure (II) obtains:

$$(C_4H_7) \vdots CH(CH_3)-CH_2-CH_3 \quad (I) \qquad (C_4H_9)-CH=CHCH_2CH_3 \quad (II)$$

m/e 55 | *m/e* 57

The former would account for the 57 peak and the base peak at *m/e* 55, by the fragmentation shown. The alternative (II) might be expected to lead to a similar spectrum. Now the observed 57 peak is not very intense ($\sim 12\%$) in the present example and probably excludes (I). Methyl branching is implied by the relatively intense peak (35%) at *m/e* 97 $[P-15]^+$. This could arise from structure (I) by loss of CH_3 at the branching point or from structure (II) if a methyl branch were present in the C_4H_9 residue. Comparison with the reported A.P.I. spectra, shows in fact, that the sample is 2,2-dimethylhex-3-ene.

There are so many octenes whose spectra are not available, that it would be most unwise to make this diagnosis on a completely unknown sample. Other evidence should be used wherever possible to help solve such problems. For example, in the present instance, the product of hydrogenation, 2,2-dimethylhexane, provides a useful point for investigation. The mass spectrum of this paraffin would at least indicate the presence of the t-butyl group, although it would not be possible to distinguish between $22M_2C_6$, $224M_3C_5$ and $2233M_4C_4$. Now since a tertiary butenyl structure is obviously impossible, we may exclude the carbon skeleton of $2233M_4C_4$ from consideration in the original octene; equally we may infer the presence of a t-butyl group in this unknown. The octene precursors of $224M_3C_5$ are 2,4,4-trimethylpent-1-ene and 2,4,4-trimethylpent-2-ene. Neither of these can lose an ethyl group to give *m/e* 83 (cf. A.P.I. spectra 130 and 131) whereas this requirement must be met for Spectrum 5 (*m/e* 82 is about 60%). Therefore the most probable olefin is 2,2-dimethylhex-3-ene, i.e.

$$CH_3-C(CH_3)_2-CH=CH-CH_2-CH_3$$

Thus, hydrogenation, leading to a knowledge of part of the carbon skeletal structure, could be of material assistance in solving this problem.

In this particular example, the impurity is readily detected since the high parent peak sensitivity of aromatic compounds makes them easy to observe even at low concentration. On the other hand, where many structural isomers exist, 92 olefins in the present example, it is often very difficult to be definite about the compound concerned.

Spectrum No. 6

The mass spectrum of this compound shows that it has a molecular weight of 88.

As mentioned in the introduction, these compounds are assumed to contain the normal isotopic distribution of the elements. It is possible, therefore, in principle, to calculate the number of atoms of each species present in the molecule. To do this, one must have the relative abundances of the parent molecular ion $[P]^+$ and the next higher masses $[P+1]^+$ and $[P+2]^+$. In this example, the relative abundances of these ions *m/e* 88, 89 and 90 are as 100 : 9·06 : 0.

Now the ratio $^{12}C:^{13}C$ is 98·9:1·11 and therefore $^{13}C/^{12}C = 0{\cdot}0112$. Assuming, as will subsequently be proved, that there are no nitrogen atoms in the molecule, the number of carbons may be determined as follows. The contributions to the $[P+1]^+$ ion from ^2H or ^{17}O may be neglected on account of the normally very low abundance of these atoms. From the Binomial theorem,

$$({}^{12}C+{}^{13}C)^n = {}^{12}C^n + n\,{}^{12}C^{n+1}\,{}^{13}C + \frac{n(n-1)}{2}\,{}^{12}C^{n+2}\,{}^{13}C^2 + \ldots$$

The second term of this expansion gives the number of ^{13}C atoms present in the molecule; since in the ion $[P]^+$ only ^{12}C atoms are present, the ion $[P+1]^+$ will represent the contribution of ^{13}C. Therefore it is necessary only to equate these to determine the value of n.

Re-arranging the above equation and substituting for the ratio $^{13}C/^{12}C$

$$\left(1+\frac{^{13}C}{^{12}C}\right)^n = (1+0{\cdot}0112)^n = 1+n\times 0{\cdot}0112+\frac{n(n-1)}{2}\times 0{\cdot}0112^2+\ldots$$

Equating,

$$[P]^+ = 1 \quad \text{and} \quad 0{\cdot}112n\,[P+1]^+ = 0{\cdot}0906,$$

whence $n = 9$.

Incidentally, the same expansion shows that, unless there is a very large number of carbon atoms, the chance of having two ^{13}C atoms in the molecule is so small as to make no significant contribution to the abundance of the $[P+2]^+$ ion.

The calculation has been carried out upon this rather unfavourable example in order to show the weakness as well as the advantage of the method. It will be shown later that the molecule does contain oxygen and, since the isotope ^{18}O makes a significant contribution to the usual natural distribution of these isotopes, $^{18}O/^{16}O = 0{\cdot}2\%$, one might expect the presence of a $[P+2]^+$ ion from which, by a comparable calculation,

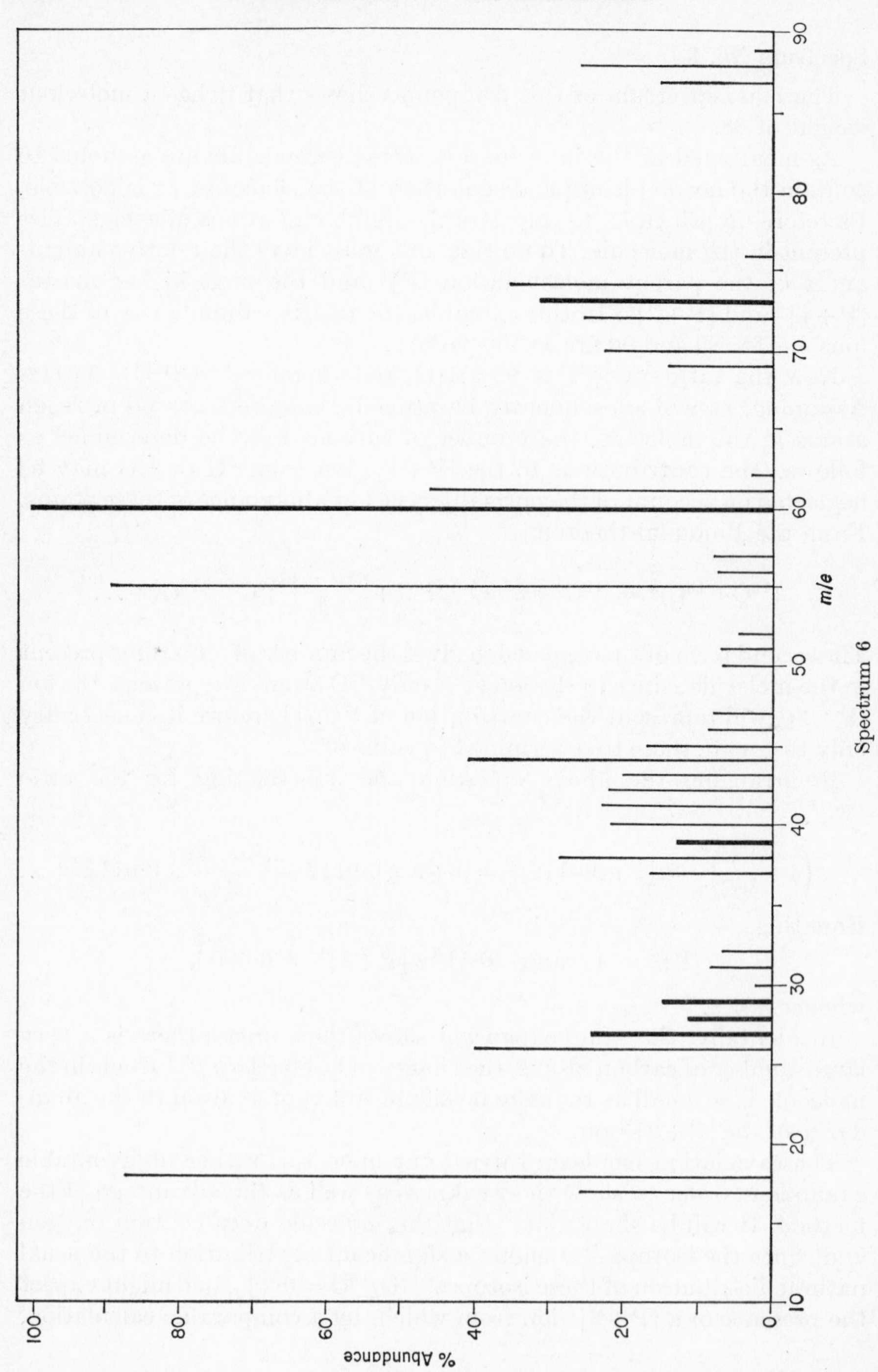

Spectrum 6

the number of oxygen atoms could be determined. Calculation shows that the abundance of $[P+2]^+$ is 0·4 relative to $[P]^+ = 100$. Since the abundance of the parent molecular ion with regard to the base peak is only 2·58, that of $[P+2]^+$ would be 0·01, too small to have been recorded on the diagram.

In this example, the abundance of the $[P+1]^+$ ion m/e 89 relative to m/e 88 is 9·06%, which would indicate the maximum number of carbon atoms to be eight or perhaps nine. This is manifestly absurd, as the molecular weight sets an upper limit of seven and the $[P+1]^+$ peak must presumably be augmented by an ion–molecule reaction (cf. Introduction, p. 2). The molecular ion must be fairly readily protonated, i.e. the molecule is basic relative to the conditions existing in an electron beam. Molecular formulae which are consistent with the known mass include, for example, C_7H_4, $C_5H_{12}O$, $C_4H_8O_2$, $C_4H_{12}N_2$ and $C_3H_8N_2O$. Some of these are readily excluded by a consideration of the cracking pattern. If the molecule is a hydrocarbon the moderately abundant fragment ion at m/e 55 must have the formula $[C_4H_7]^+$. This is inconsistent with the only hydrocarbon acceptable from consideration of the molecular weight, namely C_7H_4, and the molecule must therefore contain a hetero-atom. Nitrogen atoms are unlikely because, with the exception of the base peak at m/e 60, fragments are of *odd* mass. Unless the two nitrogens are directly joined, moreover, one would expect ions of the form $[NH_2]^+$, $[CH_2NH_2]^+$ etc., and these are not present in the spectrum. However, since the failure of most single-focusing instruments is that they do not permit the operator to decide accurately the atomic constitution of ions, the possibility that nitrogen is present cannot be entirely rejected. There are two moderately abundant ions at m/e 17 and 18 which are significant in setting the molecular formula. Taken together these represent $[OH]^+$ and $[H_2O]^+$, the relative abundances of the two peaks being consistent with this view. The molecule must contain a hydroxyl group to yield these ions and also a suitably disposed hydrogen, so that water may be eliminated. This last information is of limited use at the present stage of the analysis and, given that the molecule contains oxygen though probably not nitrogen, the molecular formula is either $C_5H_{12}O$ or $C_4H_8O_2$. If the former, it must be an alcohol. The spectra of most of the pentanols are already known (Ref. 13, Nos. 290, 291, 292, 653, 654, 655, 656, 657, 800, 801) and the present cracking pattern is not in agreement with these (cf. Spectrum No. 3). Therefore, the molecule cannot be a simple alcohol and must contain two oxygen atoms.

This argument relies upon the straightforward comparison of the spectra of known and unknown compounds. It is possible however to exclude an alcohol by other arguments. Reference to Spectrum No. 3

shows that the absence of a peak at *m/e* 31 would contra-indicate a primary alcohol. For secondary alcohols, the elimination of molecular water would leave an olefin and the ionization potential of an olefin is less than that of water; therefore, the olefin ion will be preferentially formed. This generalization seems justified since the known ionization potentials of olefins (ethylene excepted) are all less than propylene (10·02 eV) which is in turn less than water (10·10 eV). In the case of a secondary

TABLE 5

m/e	% Ab. (*I*)	$\frac{100 \times I}{\Sigma^*}$	*m/e*	% Ab. (*I*)	$\frac{100 \times I}{\Sigma}$
17	26·0	0·98	50	0·69	0·26
18	8·69	3·27	51	0·70	0·26
26	3·99	1·50	52	0·44	0·11
27	24·56	9·24	55	8·92	3·36
28	11·58	4·36	56	0·77	0·29
29	14·97	5·63	57	0·82	0·31
30	0·23	0·09	60	100·00	37·62
31	0·86	0·32	61	4·62	1·74
32	0·73	0·27	62	0·58	0·22
37	1·65	0·62	69	1·43	0·54
38	1·07	2·85	70	2·28	0·86
39	13·16	4·95	71	2·56	0·96
40	2·18	0·82	73	31·58	11·88
41	2·32	0·87	74	3·54	1·33
42	2·22	0·84	76	0·37	0·14
43	2·02	0·76	87	1·51	0·57
44	4·12	1·55	88	2·58	0·97
45	1·72	0·65	89	0·26	0·10
46	0·90	0·34			

$\Sigma = 265{\cdot}81$

* Represents the sum of the ion abundances.

pentanol, there should be an abundant ion at *m/e* 70 corresponding to the olefin formed after water has been eliminated. No such ion is present, and the molecule is most unlikely to be secondary alcohol. This argument may readily be extended to tertiary alcohols which, of course, can also be excluded on an earlier generalization, viz. that tetra-substituted molecular ions are present only to a small extent if at all. Once again we are led to the conclusion that the molecule contains two oxygen atoms and has the formula $C_4H_8O_2$. With this formula the cracking pattern may be interpreted.

Excluding the base peak, the abundant ions in the spectrum are *m/e* 73, 39, 29, 28 and 27. The first of these represents the loss of a methyl group from the parent molecular ion. The second can only have the formula $[C_3H_3]^+$ from which one can deduce that the original structure possessed a chain of at least three carbon atoms, all of which had hydrogen attached. This ion $[C_3H_3]^+$ has been discussed by several workers[15] who have concluded that it owes its stability to its "aromatic" character. The ion is often observed in spectra which contain an n-propyl or isopropyl group: the present molecule, therefore, probably contains a $\cdot C_3H_7$ grouping. The ion at *m/e* 27 can have only one formula $[C_2H_3]^+$ also. It is unlikely that this arises from further breakdown of the $[C_3H_3]^+$ ion, but it is readily derived by the following fragmentation.

$$\underset{m/e\ 29}{[C_2H_5]^+} \rightarrow \underset{m/e\ 27}{[C_2H_3]^+} + H_2$$

This reaction is, however, thermodynamically unfavourable, since the heats of formation of the ethyl and vinyl ions are 9·71 eV and 12·14 eV, respectively.[15] There is an abundant ion at *m/e* 29 which may be $[C_2H_5]^+$ although there is some ambiguity in this as $[CHO]^+$ has the same mass.

The above information allows one to deduce at least a partial structure. The molecule possesses an $-OH$, a $-CH_3$ and a $-C_3H_7$ group. There remains one oxygen to assign and since this, too, must be attached to a carbon atom, the methyl group is contained in the propyl group. That the oxygen is attached to a carbon is made very probable by the presence of the abundant ion at *m/e* 28 probably, but not unambiguously, assigned to carbon monoxide. The alternative, ethylene, is unlikely since ethylene normally appears in great abundance when it is formed, for example by elimination from a cyclic structure by the breaking of two bonds. The molecular formula $C_4H_8O_2$ has only one double bond equivalent and, if this is in a cyclic structure, the second oxygen can be present only as a further hydroxyl or as a tetrahydrofuran derivative. The former possibility would yield a glycol: the fragmentation patterns of glycols and tetrahydrofurans are very different from our present example. It is, moreover, difficult to derive the base peak *m/e* 60 from any of these molecules. The remaining possibility is that we are dealing with an acyclic molecule containing a carbonyl group. Since the molecule contains a carbonyl, a hydroxyl and a propyl group, one convenient way of arranging them is as butanoic acid C_3H_7COOH.

It now remains to determine the structure of the alkyl group which may be done as follows. The rather extensive investigations which have been carried out on the carboxylic acids[8, 16] show that a very common re-arrangement takes the following form:

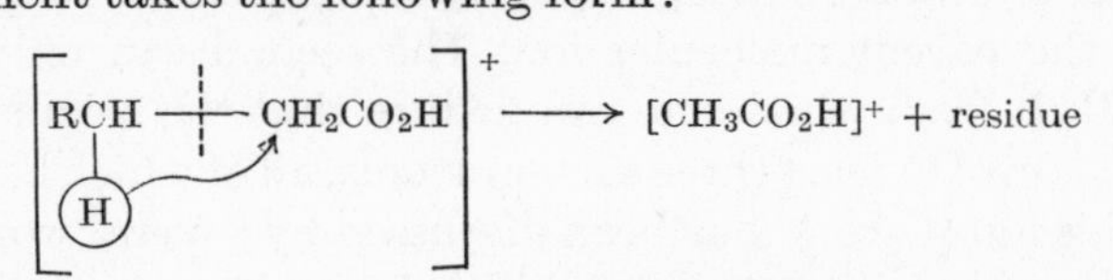

The acetic acid ion so produced would yield an ion at *m/e* 60. The β-fission with a concomitant hydrogen re-arrangement is of rather frequent occurrence in the mass spectra of compounds and, as in this case, is often useful in structure determination. The necessary condition that the re-arrangement product shall be the acetic acid ion means obviously that the carbon α to the functional group must possess two hydrogen atoms. In our present spectrum the base peak occurs at *m/e* 60 which is consistent with the ion from acetic acid. From this it follows that the partial structure of the original compound was $-CH_2CO_2H$. This excludes an isopropyl group in the molecule which must, therefore, be butanoic acid.

The identification of the compound leads to the explanation of certain properties of the spectrum which have already been noted. The abnormal abundance of the $[P+1]^+$ ion in the original mass spectrum would indicate the presence of some protonated butanoic acid $[C_3H_7CO_2H_2]^+$ in the ion.

The remaining ambiguity which may now be resolved is the high abundance of the ion at *m/e* 28. It is clear that there may be more than one ionic species at this mass, and in the present instance, both ethylene and carbon monoxide ions probably contribute to the abundance of the ion. This does not vitiate the conclusions already reached as the fission process below would not be likely to occur unless the carbonyl group were present.

$$[CH_3CH_2{-}CH_2{-}CO_2H]^+ \longrightarrow C_2H_4 + [CH_3CO_2H]^+$$
$$\text{or } [C_2H_4]^+ + CH_3CO_2H$$

The carbon monoxide ion could result from fairly extensive decomposition of the parent molecular ion only, and such extensive fragmentations and re-arrangements are known to occur.[17]

The ion at *m/e* 55 will most likely have the constitution $[C_3H_3O]^+$ presumably formed by a series of fissions within the molecular ion

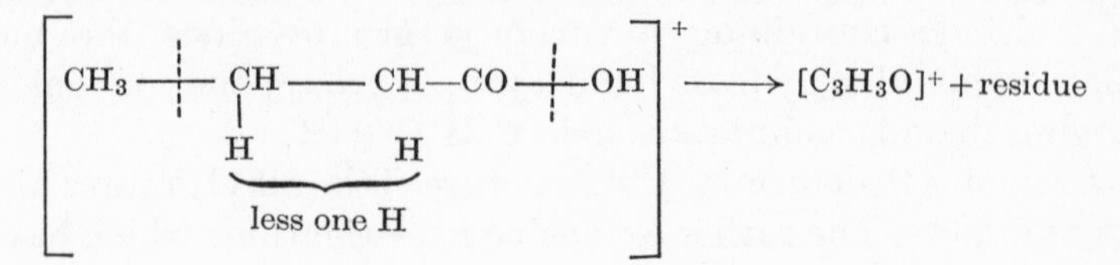

or by a further breakdown of a abundant fragment ion e.g.

$$[CH_2CH_2CO_2H]^+ \rightarrow [C_3H_3O]^+ + H_2O$$

$$m/e\ 73 \qquad\qquad m/e\ 55$$

There remain some weakly abundant ions which have not been examined. In view, however, of the possibility of re-arrangements and the difficulty of assigning an unambiguous formula, discussion of these is unprofitable.

This analysis is greatly complicated by the presence of the protonated butanoic acid which makes the determination of the molecular formula difficult.

Spectrum No. 7

The mass spectrum shows that this compound has a molecular weight of 100. The parent molecular ion is not very abundant being less than 3% of the total ion current (Table 6), and so the molecule must contain either a branched chain or a heteroatom. Possible molecular formulas will include C_7H_{16}, $C_6H_{12}O$ and $C_5H_{12}N_2$.

The abundances of the parent molecular ion $[P]^+$, the $[P+1]^+$ and the $[P+2]^+$ ions are in the ratio of 100 : 8·72 : 0·76, a result which sets an upper limit of eight carbon atoms in the molecule. This carbon content is too high: eight carbon atoms would leave so few hydrogen atoms (four) as to be quite inconsistent with the fragment ions. If the compound is a hydrocarbon, then the abundant fragment ion at *m/e* 57 must necessarily have the formula $[C_4H_9]^+$, which contains more hydrogen than was assigned to the whole molecule! The too large abundance of the $[P+1]^+$ ion may arise from two causes. Firstly the molecular ion may contain an even number of nitrogen atoms, and secondly an ion-molecule interaction may yield an ion of mass $[P+1]^+$, possibilities which are not mutually exclusive, and indeed both may make their contribution. The ratio of ^{14}N : ^{15}N is 100 : 0·381, so that even if the maximum allowable number of carbon atoms is present in the molecule, the discrepancy between the observed and calculated abundances is much too great to attribute to nitrogen. If two nitrogen atoms are present the molecular weight of the molecule sets an upper limit of five carbon atoms. The isotope contribution from the carbon atoms would then be 5·5% and, from the nitrogen, 0·76%; the total, 6·26%, is about 2·5% below the observed abundance.

For the formula C_7H_{16}, the discrepancy between the observed and calculated values is only about 1%. The presence of a significant abundance of the ion $[P+2]^+$ provides a useful lead. If we assume that the molecule contains the maximum number of carbon atoms allowed from the observed ion abundances, a molecule containing eight carbon atoms would have the ions P^+ : $[P+1]^+$: $[P+2]^+$ in the ratio 100 : 8·8 : 0·345. The calculated abundance of the ion at *m/e* 102 is about 0·4% greater than this but, since the largest permissible number of carbon atoms was used in this calculation, the discrepancy may in fact be larger. Even so, it is sufficient to suggest that there is an atom in the molecule which has an isotope of small abundances with a mass two units greater than the commonest atom. Oxygen with stable isotopes in the ratio ^{16}O : ^{17}O : ^{18}O :: 99·76 : 0·04 : 0·2 clearly meets our requirements. If two oxygen atoms are present in the molecule its formula must be $C_5H_8O_2$. The spectrum, however, cannot be interpreted on this basis. There is no evidence for hydroxyl or carboxylic acid groups such as were encountered in the

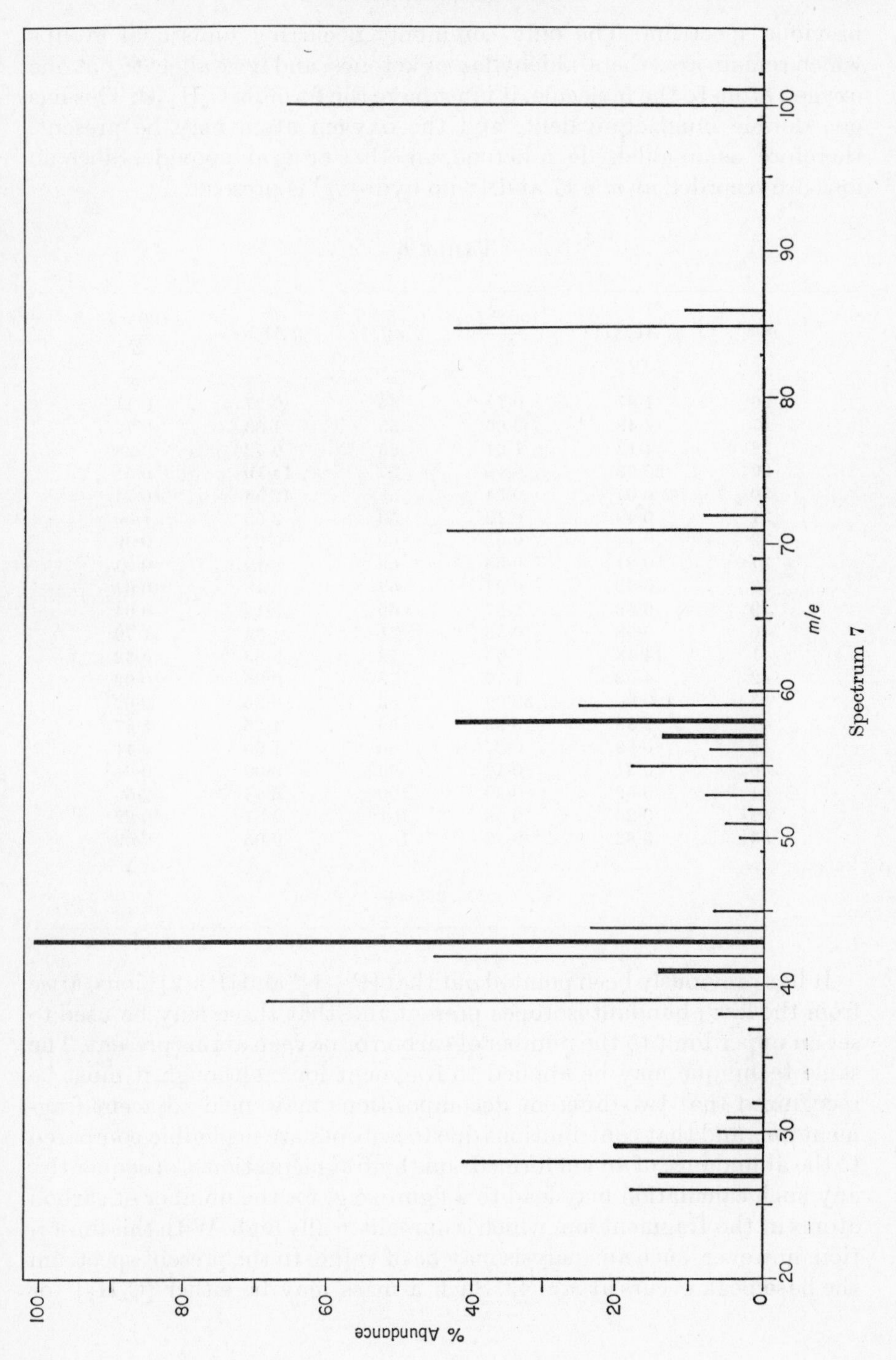

Spectrum 7

previous spectrum. The only commonly occurring functional groups which remain are ethers, aldehydes, or ketones, and if we allocate but one oxygen atom to the molecule, it must have the formula $C_6H_{12}O$. This has one double bond equivalent, and the oxygen atom may be present, therefore, as an aldehyde, a ketone, an ether or as an epoxide. Since no ions are recorded at *m/e* 17 and 18 no hydroxyl is present.

TABLE 6

m/e	% Ab. (*I*)	$\frac{100 \times I}{\Sigma}$	*m/e*	% Ab. (*I*)	$\frac{100 \times I}{\Sigma}$
26	1·81	0·71	54	0·27	0·11
27	14·48	5·66	55	1·83	0·72
28	4·13	1·61	56	0·72	0·28
29	20·93	8·18	57	14·10	5·51
30	0·61	0·24	58	42·84	16·74
31	0·49	0·19	59	2·55	1·00
32	0·14	0·05	60	0·22	0·09
37	0·21	0·08	65	0·10	0·04
38	0·62	0·24	67	0·18	0·07
39	6·82	2·67	69	0·12	0·04
40	0·92	0·36	71	4·34	1·70
41	14·48	5·66	72	0·83	0·32
42	4·52	1·77	73	0·05	0·02
43	100·00	39·09	83	0·66	0·02
44	2·38	0·93	85	4·26	1·67
45	0·69	0·27	86	1·05	0·41
50	0·31	0·12	99	0·09	0·04
51	0·52	0·20	100	6·53	2·55
52	0·20	0·08	101	0·57	0·22
53	0·82	0·32	102	0·05	0·02

$\Sigma = 255{\cdot}84$

It has previously been pointed out that $[P+1]^+$ and $[P+2]^+$ ions, arise from the less abundant isotopes present and that these may be used to set an upper limit to the number of carbon or oxygen atoms present. The same technique may be applied to fragment ions, although it must be recognized that two different decompositions may yield adjacent fragment ions and that contributions due to isotopes are negligible compared to the abundance of an ion formed thus by fragmentation. Consequently, any such calculation may lead to a figure, e.g. for the number of carbon atoms in the fragment ion, which is unrealistically high. With this limitation, however, such an analysis may be of value. In the present spectrum the base peak occurs at *m/e* 43. Such a mass may be either $[C_3H_7]^+$ or

$[C_2H_3O]^+$. The ratio of the abundances of the ions at *m/e* 43, 44, 45 is 100:2·3:0·7, a ratio consistent with a maximum of two carbon atoms and one which will permit of up to three oxygen atoms. The formula $[C_3H_7]^+$ must be excluded and the ion has the formula $[C_2H_3O]^+$. Now it is known that the acetyl group, if present in a structure, is often an abundant fragment ion (*m/e* 43). This is the base peak of the present spectrum, and the compound is thus of the form $CH_3COC_4H_9$. The molecule has only one "double bond equivalent" now identified as the acetyl group: it must therefore be an acyclic hexanone. It remains to determine the nature of the butyl group and to identify the remaining fragment ions.

The most striking property of ketones is that of β-fission with concomitant hydrogen re-arrangement.[3] In the general case of a molecule which is a methyl ketone, acetone would thus be formed and the peak at *m/e* 58 in the present spectrum may be identified with this substance. Accordingly, the original compound must have a partial formula of CH_3COCH_2CH which represents the minimum structural unit to yield acetone. The remaining C_2H_6 fragment may be arranged so that the molecule terminates either in a n- or an isopropyl group. Of these, the former is preferred because of the very abundant ion at *m/e* 29 representing either $[C_2H_5]^+$ or $[CHO]^+$. Whilst it is difficult to see how the latter ion might be derived in the present example, the formation of an ethyl ion, from 2-oxohexane, would be a straightforward fission of a single bond.

$$\left[CH_3{-}CO{-}CH_2{-}CH_2 \,\vdots\, CH_2{-}CH_3\right]^+ \longrightarrow [C_2H_5]^+ + \text{residue}$$

This would not be so for the alternative branched-chain molecule in which at least one prior re-arrangement would have to occur. Accordingly the probable structure of the compound would be $CH_3COCH_2CH_2CH_2CH_3$.

This structure may now be checked for correctness against the remaining fragment ions in the spectrum.

These occur at *m/e* 57, 41 and 27. The first of these could arise in two ways: either from the $[C_4H_9]^+$ ion formed as below (the same fission as yielded acetyl)

$$\left[CH_3CO \,\vdots\, CH_2CH_2CH_2CH_3\right]^+ \longrightarrow [C_4H_9]^+ + \text{residue}$$

or from the acetone ion which has lost one hydrogen atom.

$$\left[CH_3COCH_2 \,\vdots\, CH_2CH_2CH_3\right]^+ \longrightarrow [CH_3COCH_2]^+ + \text{residue}$$

The ion at *m/e* 41 almost certainly has the formula $[C_3H_5]^+$. This ion is often abundant in the mass spectra of compounds which contain an n-propyl group (Ref. 13, Nos. 6, 14, 39, 109, 132 *et seq.*) and is a further reason for preferring a straight chain rather than a branched structure. The ion at *m/e* 27 must be $[C_2H_3]^+$, often found in alkane spectra (Ref. 13, as above) in association with the $[C_2H_5]^+$ as is the case here.

In addition to all the abundant ions already listed, there are some of moderate abundance at *m/e* 85, 71, 59, 42, 39 and 28. These, in order, may now be assigned the formulae $[C_5H_9O]^+$, $[C_4H_7O]^+$, $[C_3H_7O]^+$, $[C_3H_6]^+$, $[C_3H_3]^+$ and $[CO]^+$. Three of these are unambiguous structures; namely $[C_5H_9O]^+$ which refers to the loss of a methyl group from the parent molecular ion; $[C_3H_7O]^+$ which must be present in the high mass satellite of the acetone ion and is probably protonated acetone; and $[C_3H_3]^+$ (cf. Spectrum No. 6, p. 151) which is often observed when there is the requisite number of carbon atoms joined to each other. The ion $[C_4H_7O]^+$ probably results when an ethyl group is broken off and is further evidence in support of the proposed structure. The ion at *m/e* 55 cannot be given a definite derivation nor can a single formula be assigned to it: two possibilities, both consistent with the mass, are $[C_4H_7]^+$ and $[C_3H_3O]^+$. Each could be derived from the known structure, the former by removal of two hydrogen atoms from the butyl ion, and the latter either by extensive loss of hydrogen from the acetone ion, or more likely by the loss of molecular hydrogen from the other possible ion of *m/e* 57, viz. $[CH_3COCH_2]^+$. The ion at *m/e* 42 seems likely to be the propylene ion formed along with acetone during the hydrogen re-arrangement.

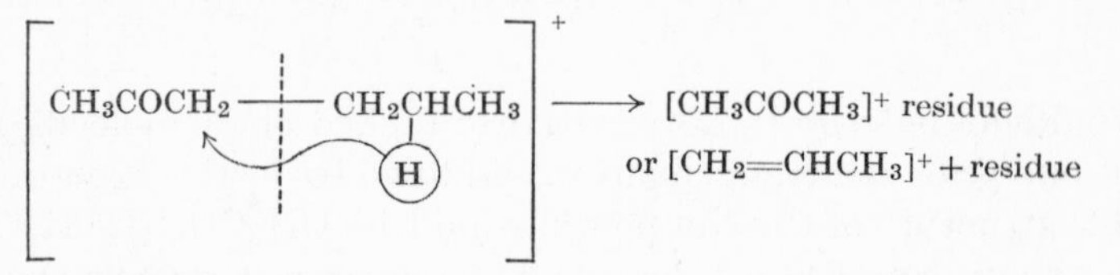

The remaining ion (*m/e* 28) may refer either to $[CO]^+$ or $[C_2H_4]^+$; both probably contribute.

This analysis is greatly facilitated, as indeed was the preceding one, by a knowledge of the elements contained in the molecule. Since no nitrogen was present, the abundant ion of even mass must be a molecular ion for reasons already discussed in the introduction. Had the next higher peak been in the correct ratio with that of acetone namely *m/e* 58 : *m/e* 59 as is 100 : 3·3 instead of the observed ratio of 100 : 5·4, there would have been no difficulty in recognizing the process that had occurred, and much of the argument here detailed would have been irrelevant. This failure to distinguish ions of the same nominal mass, but having small mass differences, is perhaps the great disadvantage of single-focusing instruments.

Spectrum No. 8

Preliminary information supplied is that the substance has m.p. 83–84° C.

The mass spectrum of this compound shows a doublet giving two parent molecular ions at *m/e* 157 and 159. These are present in the ratio 3:1 which is the same as that between the naturally occurring isotopes, ^{35}Cl and ^{37}Cl. Accordingly, the molecule must contain one atom of chlorine. The molecular weight of each species, moreover, is an odd number, which in turn indicates that an odd number of nitrogen atoms is present. Corresponding to each parent molecular ion, there is a $[P+1]^+$ and a $[P+2]^+$ ion which enables us to set an upper limit to the number of carbon and oxygen atoms in the molecule. The contribution of the oxygen isotope in the molecule with ^{35}Cl is obscured by the main ion of the ^{37}Cl molecule, but the information may be obtained from the ion at 161. This relates to the molecules having ^{37}Cl; it is not obscured by the other species. For the series *m/e* 159, 160, 161, the abundances are in the ratio 100:6:0·45. The molecule, therefore, can contain a maximum of five or six carbon atoms and two oxygen atoms in addition to the chlorine and the odd number of nitrogens already found. This molecule clearly has a cyclic structure, since the molecular ion, which contains the most abundant chlorine isotope, is also the base peak of the spectrum. Furthermore, the chlorine atom remains firmly attached to the structure, as is shown by the recognizable chlorine-containing doublets at *m/e* 113 and 111, at 101 and 99, and at 52 and 50, a view confirmed by the absence of abundant ions at *m/e* 35 and 37 or at *m/e* 36 and 38 which would refer to chlorine and hydrogen chloride ions respectively.

Further discussion will be greatly simplified if a monoisotopic spectrum is considered, and unless specifically contradicted, the remainder of the discussion will be concerned only with the molecules which contain ^{35}Cl.

Since the molecular weight is 157, and the molecule probably contains two oxygens, five or six carbons and an odd number of nitrogen atoms, possible formulae include $C_6H_4NO_2{}^{35}Cl$, $C_5H_4N_3O{}^{35}Cl$, etc. A choice amongst these possibilities is not easily made since there are few abundant ions which may be assigned a single chemical formula. The most suitable peak to start on is that at *m/e* 76. It is not part of a doublet and is therefore free from chlorine. Valency requirements of the atoms in this compound are such that this ion may be $[C_6H_4]^+$ or $[C_5H_2N]^+$. Since these ions contain so little hydrogen compared to carbon, they are highly unsaturated systems. Linear molecules must necessarily contain acetylene units and such molecules are known to fragment rather easily (Ref. 13 Nos. 38, 532–537 inclusive). Unsaturated cyclic structures derived

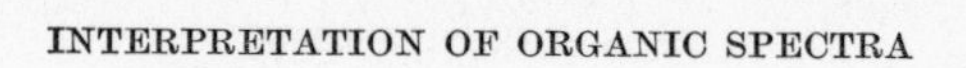

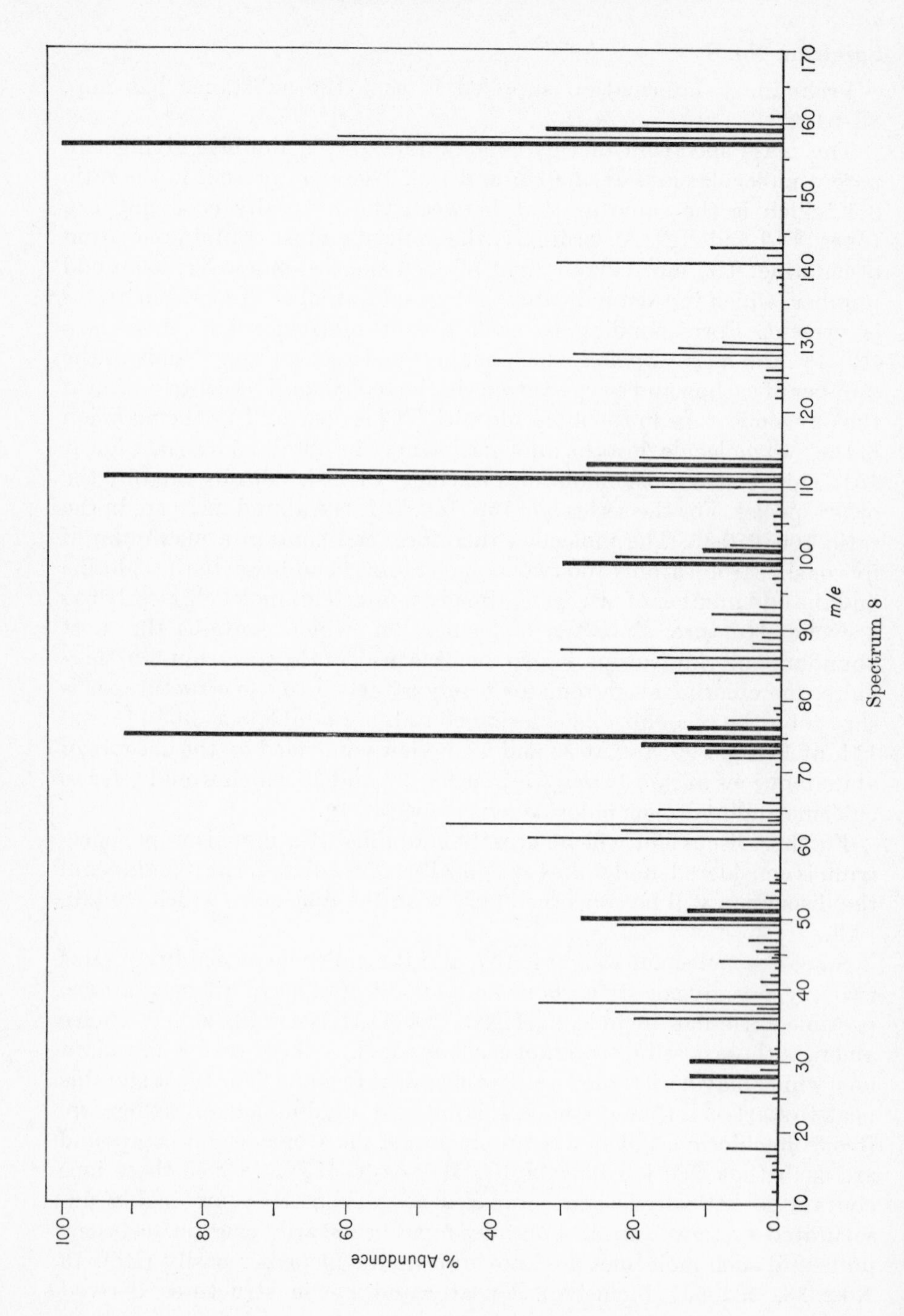

% Abundance
100
80
60
40
20
0
10
20
30
40
50
60
70
80
90
100
110
120
130
140
150
160
170
m/e

Spectrum 8

from benzene or pyridine, on the other hand, produce ions of considerable stability, and the presence of the ion at *m/e* 76 suggests that we are dealing with this type of structure. In discussions of previous spectra, it has already been noted that the $[P+1]^+$ ion can have a greater abundance than would be deduced from isotope considerations and that this might arise from protonation (compare the four molecular ions of acetic and butanoic acids, of acetone and 2-oxohexane). Now pyridine is clearly more basic than any of these and thus one might expect a significant contribution to the spectrum from the pyridinium ion. In the present spectrum, however, there is no abnormally large $[P+1]^+$ ion, and the molecule therefore is neither a substituted pyridine nor an aminobenzene to which the same arguments would apply; we are dealing rather with a substituted benzene in which the nitrogen is not part of a basic group. The molecular formula must be $C_6H_4NO_2{}^{35}Cl$; the chlorine, firmly attached to the carbon skeleton, must be directly joined to the benzene ring.

In addition to the parent molecular ion, there is a series of abundant fragment ions (F) at *m/e* 111, 99, 76, 75, 74, 73, 51, 30 and 28. The first represents the loss of 46 mass units from the parent molecular ion and the abundance of this ion at *m/e* 111 relative to those at $[F+1]^+$ and $[F+2]^+$ gives the ratio of 100 : 6·67 : 0·2. This sets an upper limit of six carbon atoms and one oxygen atom, and requires that at least one oxygen is included in the 46 mass units lost. The fragment eliminated must have the formula CH_4NO, CH_2O_2 or NO_2 of which the last is the most reasonable since neither of the first two fragments could be removed to leave a reasonably stable residual ion. The molecule is, therefore, a nitrochlorobenzene. The ion at *m/e* 99 represents the loss of a further 12 units and, having regard to the abundance of *m/e* 100, means the elimination from the parent molecular ion of CNO_2. Additional information about the nature of this elimination will be considered in the subsequent discussion of moderately abundant ions. The next ion at *m/e* 76 has been referred to already and was used to determine the molecular formula. The structure corresponding to the very abundant peak *m/e* 75 can be only $[C_6H_3]^+$, which represents a further loss of hydrogen from the preceeding ion, although it need not be obtained directly in this way. The ion of the next lower mass *m/e* 74 may represent the still further loss of hydrogen, while the final one of this group at *m/e* 73 probably has the formula $[C_2H_3NO_2]^+$, being formed by fragmentations across the benzene ring with concomitant hydrogen migration. The abundant ions at *m/e* 51 and 50 are also hydrocarbon ions since no other combinations of the available atoms can give the correct masses. These ions are $[C_4H_3]^+$ $[C_4H_2]^+$, respectively, and again are obtained by breakdown of the aromatic ring. Finally, there remain the two abundant ions at *m/e* 28 and 30. As to the former, it is unlikely that this ion is $[C_2H_4]^+$ since the original compound contained only four hydrogen atoms, and the ion is almost certainly due to CO^+. This conclusion will be strengthened in the subsequent discussion. The ion of mass 30 must be NO, as no other combination of the available atoms will give the correct value.

In addition to the abundant ions so far examined, there is an interesting series of moderately abundant ions which, omitting the isotope peaks, are at *m/e* 141, 128, 127, 110, 85, 63, 61 and 38. The first of these refers to the loss of a single oxygen atom from the parent, a rather unusual occurrence, but known to happen with aromatic nitro compounds under certain conditions (Ref. 3, p. 406). The next two refer to the loss of 29 and 30 mass units from the molecular ion. The neutral fragments removed must have the formulae CHO and NO, respectively, and can be eliminated only by some re-arrangement in the molecular ion; this probably accounts for their moderate abundance.

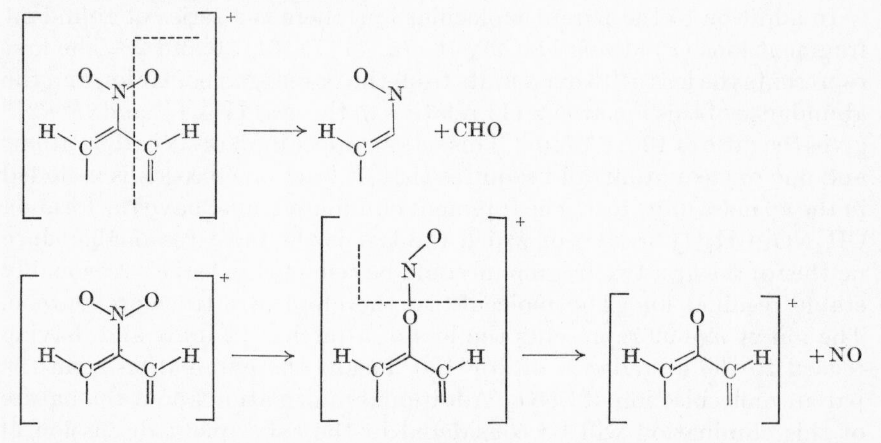

The ion at *m/e* 110 may represent the loss of HNO_2 and that of mass 85, the most abundant of this series, the removal of $C_2H_2NO_2$, as the composition of the ion must be $[C_4H_2{}^{35}Cl]^+$. There remains two ions of uncertain constitution at *m/e* 63 and 61: no information as to their composition can be obtained from isotope abundance. Finally, the ion at *m/e* 38 must have the formula $[C_3H_2]^+$ and is of considerable help in determining the orientation of the nitro and chloro substituents. So far we have deduced that we are dealing with one of the isomeric nitrochlorobenzenes. It has been noted in previous spectra that where there are three carbon atoms in a chain with an adequate supply of hydrogen atoms attached thereto, the ion at *m/e* 39 $[C_3H_3]^+$ is present in the spectrum. In the compound under discussion, the corresponding ion appears, however, at *m/e* 38, which is consistent with the conclusion that nowhere in the molecule are there three contiguous carbon atoms each possessing one hydrogen atom. The only orientation to which this restriction can apply is *p*-chloronitrobenzene. Fission of this molecule into three carbon fragments could produce the ion $[C_3H_2Cl]^+$. There would then be a doublet at *m/e* 73 and 75. Note that these ions, if present, would occur at the same nominal masses as would $[C_2H_3NO_2]^+$ and $[C_6H_3]^+$, already discussed. Alternatively, the ion $[C_3H_2NO_2]^+$ could be obtained. This is of *m/e* 84, and there is a weak ion of this mass present in the spectrum.

The most obvious piece of confirmatory evidence, however, owes nothing to mass spectrometry but illustrates the importance of not ignoring simple properties. The material used in the analysis was a solid: the only solid isomer is *para*-nitrochlorobenzene!

When in the introduction it was stated that the only evidence one could obtain was the mass and abundance of ions, this represented an over-simplification of the situation. The method of operation of a mass spectrometer is to ionize a molecule by electron bombardment, to expel it from the ion source by an electrostatic repulsion and to deflect the ions so obtained by means of a magnetic field. This has the effect of resolving the beam into a series of discrete ion beams which are homogeneous in that the mass-to-charge ratio within each is constant. If, however, an ion of mass M^+ which has been electrostatically repelled, dissociates into a smaller mass m^+ before magnetic deflection, the resulting beam is a somewhat diffuse one of low abundance. Usually it is not sharply defined and the maximum occurs at a non-integral value. The mass-to-charge ratio of this "metastable ion" is given by the formula m^2/M. Provided that we know the value of the metastable ion, the values of M and m can be determined,[18] and hence the mass of the fragment removed. In the present spectrum, there is a metastable ion at *m/e* 77·2 which corresponds to the transition $127^+ \rightarrow 99^+ + 28$. Now the structure of the ion at *m/e* 127 has already been deduced as $[C_6H_4O^{35}Cl]^+$ and, having regard to the few hydrogens available, the loss of 28 mass units as a single entity must refer to the further elimination of carbon monoxide. The ion at *m/e* 99 therefore has the formula $[C_5H_4{}^{35}Cl]^+$ already discussed. From a consideration of the known cracking patterns of aromatic quinones[129] and ethers,[20] it could conceivably be of the form of a chlorinated fulvene ion.

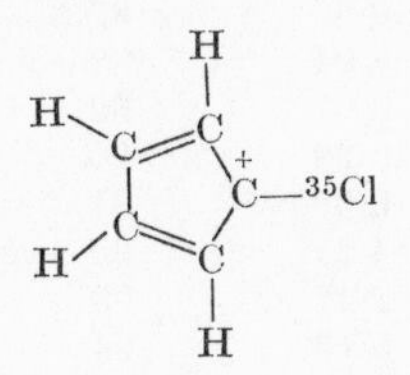

The mass spectra of molecules such as *p*-nitrochlorobenzene are often too difficult to analyse. The large amount of carbon relative to the hydrogen atoms together with the great stability of the aromatic ring are a very useful guide. Except when a basic group is present, the isotope peaks also provide good evidence of the correctness of the molecular formula and often of the composition of the more abundant fragment ions. The main difficulty comes in the determination of orientation, especially where a series of fragment ions from different fissions overlap. Moreover, simple differences in physical properties are not always available as they were on this occasion.

TABLE 7

m/e	% Ab. (*I*)	$\frac{100 \times I}{\Sigma}$	*m/e*	% Ab. (*I*)	$\frac{100 \times I}{\Sigma}$
12	0·06	0·01	53	0·21	0·04
14	0·06	0·01	54	0·05	0·01
15	0·06	0·01	55	0·11	0·02
16	0·07	0·01	55·5	0·03	0·01
17	0·15	0·03	56	0·04	0·01
18	0·71	12	60	0·47	0·08
25	0·06	0·01	61	3·52	0·60
26	0·52	0·09	62	1·92	0·33
27	1·56	0·27	63	4·67	0·80
28	12·43	2·12	63·5	0·18	0·03
29	0·24	0·04	64	1·81	0·31
30	12·14	2·07	65	1·66	0·28
31	0·11	0·02	66	0·22	0·04
32	0·05	0·01	67	0·06	0·01
35	0·30	0·05	71	0·05	0·01
36	2·03	0·35	72	0·71	0·12
37	2·24	0·38	73	10·30	1·76
37·5	0·26	0·04	74	19·18	3·27
38	3·02	0·52	75	91·59	15·62
39	1·22	0·21	76	13·26	2·26
40	0·16	0·03	77	0·73	0·12
41	0·21	0·04	78	0·06	0·01
42	0·09	0·02	79	0·09	0·02
43	0·04	0·01	83	0·06	0·01
44	0·85	0·14	84	1·45	0·25
44·5	0·09	0·02	85	8·88	1·51
45	0·06	0·01	86	1·74	0·30
45·5	0·30	0·05	87	3·05	0·52
46	0·22	0·04	88	0·37	0·06
47	0·41	0·07	89	0·25	0·04
48	1·70	0·29	90	0·94	0·16
49	2·26	0·39	91	0·29	0·05
50	24·87	4·24	92	0·98	0·67
51	14·21	2·42	93	0·30	0·05
52	0·79	0·13	94	0·04	0·01

TABLE 7—*continued*

m/e	% Ab. (I)	$\frac{100 \times I}{\Sigma}$	m/e	% Ab. (I)	$\frac{100 \times I}{\Sigma}$
96	0·05	0·01	123	0·04	0·01
97	0·12	0·02	124	0·05	0·01
98	0·27	0·05	125	0·54	0·34
99	30·49	5·20	126	0·20	0·03
100	2·25	0·38	127	4·02	0·69
101	10·95	1·87	128	2·88	0·49
102	0·69	0·12	129	1·44	0·25
103	0·04	0·01	130	0·82	0·14
104	0·06	0·01	131	0·06	0·01
105	0·09	0·02	137	0·05	0·01
106	0·08	0·01	138	0·04	0·01
107	0·04	0·01	139	0·07	0·01
108	0·59	0·10	140	0·05	0·01
109	0·44	0·08	141	3·14	0·54
110	1·81	0·31	142	0·21	0·04
111	94·08	16·05	143	0·98	0·17
112	63·8	1·07	144	0·06	0·01
113	27·24	4·65	157	100·00	17·06
114	1·97	0·34	158	6·16	1·05
115	0·05	0·01	159	32·86	5·60
122	0·09	0·02	160	1·96	0·33
			161	0·15	0·03

$\Sigma = 586{\cdot}28$

Spectrum No. 9

The cracking pattern has an abundant ion at *m/e* 104 corresponding to the parent molecular ion; isotope peaks appear on the high mass side of it. Applying the usual analysis (p. 147) to the $100 \times [P+1]^+/P^+$ ratio, the upper limit to the number of carbon atoms in this compound is five. The intensity of the peak at $[P+2]^+$ is absurdly large for this to arise from an ^{18}O isotope and either there is a significant contribution to this mass from an ion–molecule interaction, or there is an element present not so far encountered in these analyses. The ratio $100 \times [P+2]^+/P^+$ is 4·4 and the relative ratios of $P^+ : [P+1]^+ : [P+2]^+$ are 100:5·6:4·4 of which the value for the $[P+1]^+$ ion is too large by an unknown ^{13}C contribution. The corresponding figures for $^{32}S : ^{33}S : ^{34}S$ are 100:0·78:4·42; they agree well as regards the $[P+2]^+$ ion. Accordingly, it may be assumed, tentatively, that the molecule contains one sulphur atom, and a final decision on the molecular formula is most rapidly reached from some of the abundant fragment ions. The molecule cannot contain an odd number of nitrogen atoms since the molecular weight is even. If it contained two nitrogen atoms, the combined weight of nitrogen and sulphur would come to 60 leaving only 44 mass units unaccounted for. There is an ion at *m/e* 39 which must be $[C_3H_3]^+$ and, to accommodate the remaining five mass units, one possible formula would be $C_3H_8N_2S$. With this, however, it is not possible to intepret the ion at *m/e* 55. Consequently, this formula must be rejected. An alternative hypothesis, that the molecule contains oxygen, is likewise inconsistent with the observed spectrum. Since the abundance of the isotope at $[P+2]^+$ is exactly that required for one sulphur atom, we may proceed on this basis. Abundance ratios provide an upper limit to the possible number of atoms present, so we may rule out the presence of oxygen. The molecule can contain only carbon, hydrogen and sulphur and the formula must be $C_5H_{12}S$.

The next problem is to deduce the position of the sulphur atom in the carbon chain. From the formula, it is clear that the molecule contains neither a double bond nor ring structure, and the sulphur must be present as a thiol or a thio-ether. A very abundant ion at *m/e* 70 represents the loss of 34 mass units. This must be hydrogen sulphide, and since the ion is very abundant (~35%), it is not likely to arise by a double fission and re-arrangement of the type below:

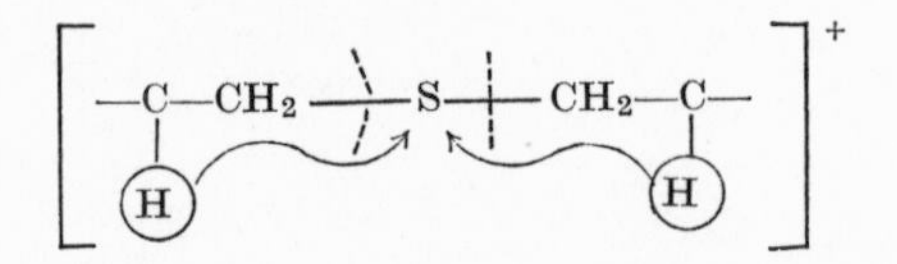

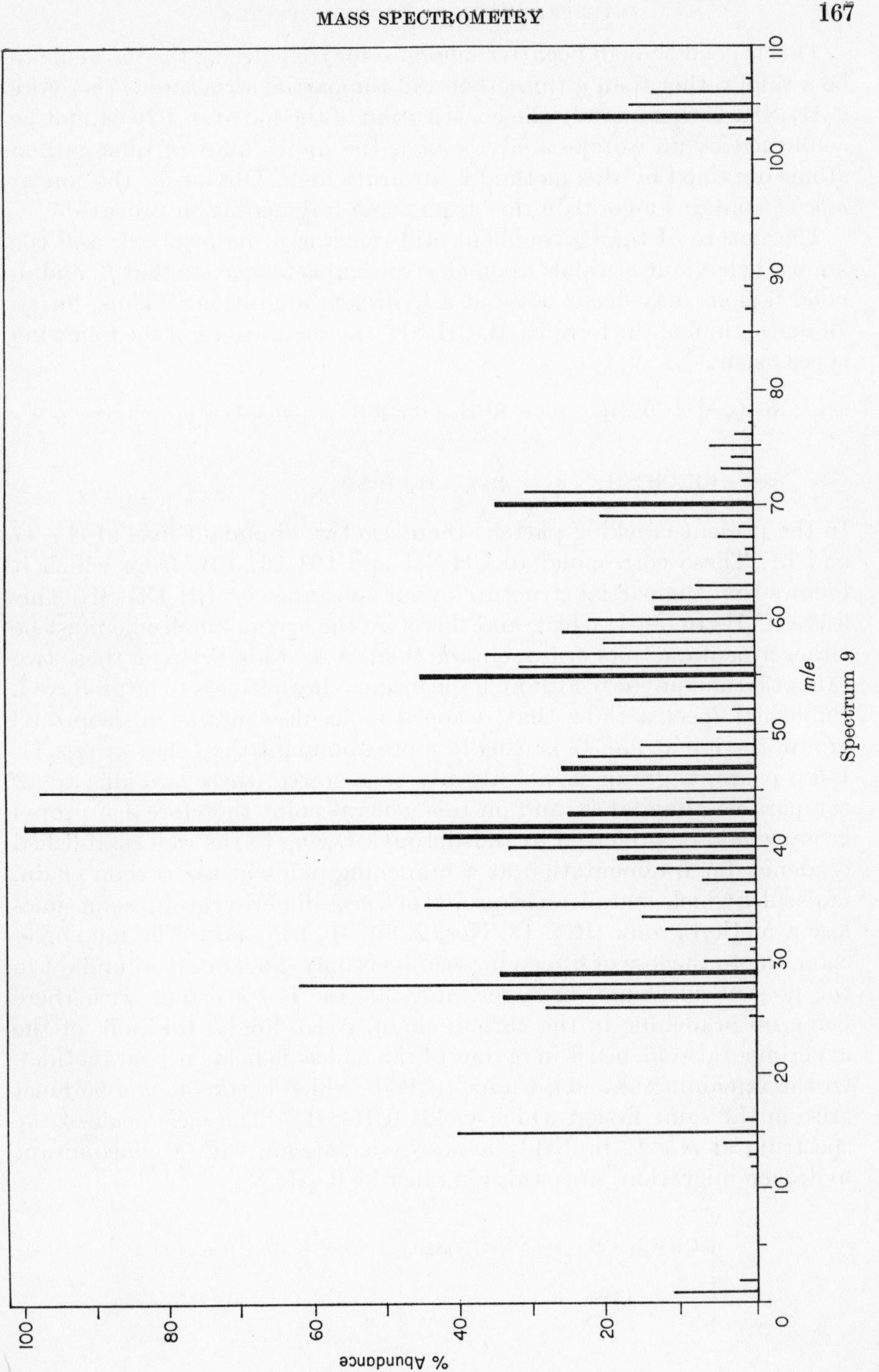

Spectrum 9

This is good though negative evidence for considering the molecule to be a thiol rather than a thio-ether and the partial structure is therefore $C_5H_{11}SH$. Unfortunately the constitution of the ion at *m/e* 70 cannot be confirmed by an isotope analysis since the upper limit of nine carbon atoms obtained by this method is absurdly high. Obviously the ion at *m/e* 71 contains a contribution from some fragmentation process.

The nature of the C_5 fragment still remains to be resolved, and the known behaviour of thiols upon electron impact suggests that β- and γ-bond fissions may occur without a hydrogen migration.[21] Thus, for an aliphatic thiol of the form RCH_2CH_2SH, fragmentations of the following types occur.

$$RCH_2 \mid CH_2SH \longrightarrow RCH_2 + [CH_2SH]^+ \text{ (or an isomer in each case)}$$

$$R \mid CH_2CH_2SH \qquad R\cdot + [CH_2CH_2SH]^+$$

In the present cracking pattern there are two abundant ions of *m/e* 47 and 61. These correspond to CH_2SH and CH_2CH_2SH, from which it follows that the partial structure of our substance is $-CH_2CH_2SH$. This leaves C_3H_7 to be attached, and therefore the original molecule must be either n-pentane thiol or isopentane thiol. A decision between these two cannot be unequivocal, although the former alternative is to be preferred. Published spectra show that, when a molecule contains an isopropyl group, the ion at *m/e* 43 is usually more abundant than that at *m/e* 41. For a n-propyl group, as has already been stated, these two ions are of comparable abundance, and on this general point therefore a n-propyl group would be preferred to an isopropyl. Owing to the well established tendency for fragmentation at a branching point in the carbon chain, molecules which contain an isopropyl or a *gem*-dimethyl group sometimes lose a methyl group (Ref. 13, Nos. 2, 15, 40, 245, 479). The ion corresponding to the loss of fifteen mass units is only moderately abundant in the present spectrum, however, and this too is consistent with there being no branching in the carbon chain. Accordingly, the bulk of the experimental evidence is in favour of the molecule being n-pentanethiol. Of the remaining abundant ions, $[C_4H_7]^+$ which occurs at *m/e* 55 must arise in the same fission which yields $[CH_2SH]^+$. The base peak of the spectrum at *m/e* 42 probably arises by a γ-fission with a concomitant hydrogen migration, since this ion must be $[C_3H_6\cdot]^+$.

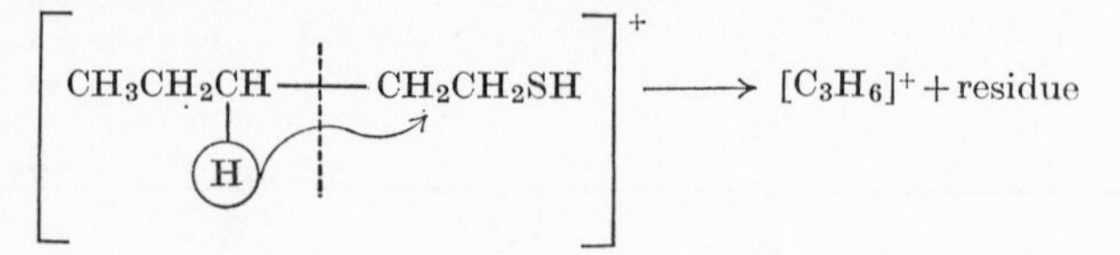

There is no abundant ion at *m/e* 62, which suggests that the sulphur-containing fragment further dissociates, and the alternative ionization to yield $[C_2H_6S]^+$ does not occur.

The two remaining abundant ions which have not been mentioned previously are at *m/e* 27 and 29. These must be the vinyl and ethyl ions. The latter may arise from n-pentanethiol by a single fission.

$$\left[C_2H_5 \vdots (CH_2)_3SH\right]^+ \longrightarrow [C_2H_5]^+ + \text{residue}$$

The vinyl ion can arise either by direct loss of hydrogen from the ethyl ion or by a separate fragmentation of the parent molecular ion. Production of an ethyl ion from a n-pentanethiol is obviously easier than from the isopentanethiol in which either a complicated re-arrangement of the terminal isopropyl group, or a double fission of a section of the pentane chain with a concomitant hydrogen migration would be necessary. While exceptions are known, (Ref. 19, p. 263), it is nevertheless true in general that ions which can arise only by multiple fissions and/or by multiple re-arrangements are never very abundant in the mass spectrum of compounds. The presence of an abundant ethyl may be taken as further proof that the correct structure has been assigned.

In addition to these very abundant ions, there are several ions of moderate abundance some of which have already been mentioned in the discussion. The ions at *m/e* 46 and 45 must both contain the hetero-atom, and have the formulae $[CH_2S]^+$ and $[CHS]^+$. The former of these would be analogous to formaldehyde, the latter to a formyl ion. These ions can be obtained only by considerable hydrogen loss, and possibly a re-arrangement, of the terminal group of the thiol, e.g.

$$[R-CH_2-SH]^+ \rightarrow R\cdot + H_2 + [CHS]^+$$

and this may be the reason for their low abundance. Again, the ion at *m/e* 35 can only be $[H_3S]^+$. It is known that H_2S is readily eliminated from the parent molecular ion, but to decide whether the above ion is formed by protonation of the hydrogen sulphide or is a fragment from an ion molecule reaction would require a knowledge of the dependance of peak height on sample pressure. The final moderately abundant ions of interest are at *m/e* 28 and 26, corresponding to $[C_2H_4]^+$ and $[C_2H_2]^+$. Both are of low abundance and, therefore, it is not possible to decide their origins.

The analysis of this compound is greatly hindered by the fact that there are a great many fragment ions present which make any isotope analysis of the most abundant ions an unrewarding exercise. Once the main outline of the molecular structure has been determined, they are of some use. These ions, which must arise by different fragmentions of the main structures with hydrogen re-arrangements, are plainly consistent with the presence of many hydrogen atoms in the molecule. In our present example, however, valency requirements and the nature of the sole functional group determine the hydrogen content of the compound, and it is not necessary to carry out any further analysis.

TABLE 8

m/e	% Ab. (I)	$\frac{100 \times I}{\Sigma}$	m/e	% Ab. (I)	$\frac{100 \times I}{\Sigma}$
1	1·14	0·23	53	1·88	0·38
2	0·23	0·05	54	0·93	0·19
12	0·06	0·01	55	45·72	9·20
13	0·10	0·02	56	3·19	0·64
14	1·08	0·22	57	3·64	0·73
15	4·10	0·82	58	2·07	0·42
16	0·15	0·03	59	2·61	0·52
25	0·08	0·02	60	2·27	0·46
26	2·87	0·58	61	13·65	2·75
27	34·45	6·93	62	1·38	0·28
28	6·25	1·26	63	0·75	0·15
29	34·96	7·03	64	0·10	0·02
30	7·40	1·49	65	0·21	0·04
31	0·08	0·02	66	0·12	0·02
32	0·39	0·08	67	0·52	0·10
33	1·26	0·25	68	0·22	0·04
34	1·56	0·31	69	2·12	0·43
35	4·53	0·91	70	35·61	7·16
36	0·16	0·03	71	3·12	0·63
37	0·75	0·15	72	0·12	0·02
38	1·35	0·27	73	0·46	0·09
39	18·67	3·76	74	0·30	0·06
40	3·34	0·67	75	0·59	0·12
41	42·50	8·55	76	0·24	0·05
42	100·00	20·11	77	0·05	0·01
43	25·37	5·10	78	0·05	0·01
44	0·95	0·19	85	0·12	0·02
45	9·78	1·97	87	0·09	0·02
46	5·22	1·05	89	0·10	0·02
47	26·27	5·28	102	0·06	0·01
48	2·40	0·48	103	0·32	0·06
49	1·51	0·30	104	30·84	6·20
50	0·55	0·11	105	1·69	0·34
51	0·80	0·16	106	1·38	0·28
52	0·30	0·06	107	0·06	0·01

$\Sigma = 497{\cdot}19$

Spectrum No. 10

The mass spectrum shows an ion of low abundance at *m/e* 101. If this is the parent molecular ion, the low abundance indicates that the structure is either greatly branched or that it contains one or more heteroatoms; the odd molecular weight requires an odd number of nitrogen atoms. When the parent molecular ion $[P]^+$ is of a very low abundance or is entirely absent, no information concerning the number or nature of the atoms present can be obtained from the $[P+1]^+$ and $[P+2]^+$ ions. In the present spectrum, therefore, the analytical sequence adopted in the four preceeding examples must be modified and more consideration given to fragment ions.

The problem of determining the molecular constitution would be almost impossible if the only information available were the above spectrum. It is aided in this particular example by examining the largest

TABLE 9

Compound	*m/e*	Compound	*m/e*
$C_3H_7N_3O$	101·091	$C_3H_9N_4$	101·115
$C_5H_9O_2$	101·093	$C_5H_{11}NO$	101·116
$C_4H_9N_2O$	101·104	$C_4H_{11}N_3$	101·128
Unknown	101·112	$C_6H_{13}O$	101·129

recorded ion upon a double-focusing mass spectrometer. In these instruments, the ions formed by electron bombardment are deflected and focused electrostatically before entering the magnetic field. The extra focusing of the ions reduces the aberrations in the ion optics and results in an enormously increased resolving power. Under these conditions, the mass of the ion may be measured with such precision that the mass defects (the small differences between the actual mass of an atom and the nearest whole number) may be used to determine molecular constitution.

In Table 9 the mass of this molecule is compared with that of a series of known constitution.

From these figures the most probable composition of the unknown is either $C_3H_9N_4$ or $C_5H_{11}NO$.

The above illustrates one of the great advantages of a double-focusing instrument namely in restricting the choice of the probable chemical composition, even where the ion of greatest mass is of such a low abundance. To confirm the composition of the compound, it is now necessary to determine the presence of oxygen when the formula will be $C_5H_{11}NO$.

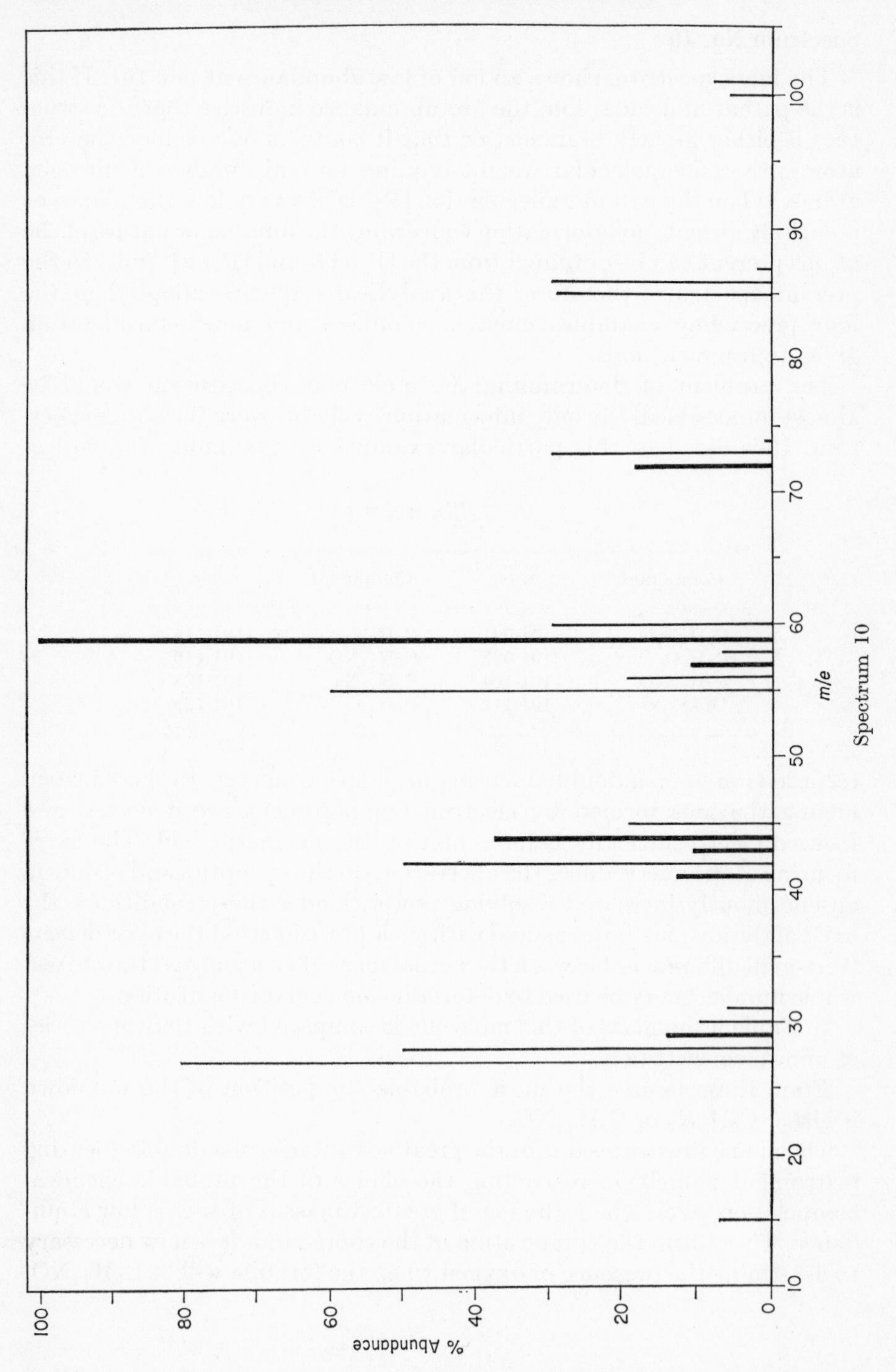

Spectrum 10

As a general principle, it is useful to concentrate initially upon the more abundant fragment ions which have an unambiguous chemical formula. If none such exist, it is most useful to examine those with the fewest possible alternatives. The larger the mass of the fragment ion which can be assigned to definite formula, the greater the chance of a successful analysis. Even if this is not possible, as in the present example, then the unknown structure may be obtained by examining possible structures in the light of their predicted breakdown patterns. In our case it is sufficient to show that the formula $C_5H_{11}NO$ is more likely than $C_3H_9N_4$ which, incidentally, would be a fragment ion.

TABLE 10

m/e	Molecular ions					
29	$C_2H_5^+$	CHO^+	CH_3N^+	N_2H^+			
72		$C_4H_{10}N$	$C_4H_8O^+$	C_3H_6NO	$C_3H_8N_2^+$		
41	$C_3H_5^+$				$C_2H_3N^+$		
59		$C_3H_7O^+$	$C_2H_7N_2^+$		$C_3H_9N^+$	$CH_5N_3^+$	$C_2H_5NO^+$
44	$C_2H_6N^+$	$CH_4N_2^+$	CH_2NO^+		$C_3H_8^+$	$C_2H_4O^+$	
57	$C_4H_9^+$	$C_2H_5N_2^+$	$C_3H_5O^+$		$C_3H_7N^+$	$C_2H_3NO^+$	

Now, when the molecule AB is ionized, the ion so formed may dissociate in two ways to yield the ions A^+ and B^+ although not necessarily in equal abundances. If A and B represent radicals in a polyatomic molecule rather than single atoms, then either or both of the ions may be further degraded particularly if hydrogen atoms are present. If pairs of such ions are present and if their sum approximates closely to the supposed molecular weight, then they are likely to be fragments of the original compound. Only pairs of ions of reasonable abundance, i.e. more than 10% of the base peak, should be considered, as ions often arise in low abundance as the result of complicated re-arrangements. Table 10 summarizes some such pairs of ions in the spectrum under discussion together with probable formulae.

The choice of the two ions used to construct the molecular formula in this case will be subject to the further requirement that the molecule so constructed shall contain not more than one oxygen atom. In the particular case of the ions *m/e* 41 and 59, it is unlikely that the original molecule would fragment to give two molecular ions and a single neutral hydrogen, and, since this is the only way by which the ion $[C_3H_9N_4]^+$ could be obtained, this formulation is unlikely. Accordingly, the alternative formula $C_5H_{11}NO$ is to be preferred.

Having chosen the molecular formula, which shows that the compound has one double-bond equivalent, it remains to determine the nature of the functional group or groups. Again the method is a tentative one, the correctness of each decision is checked by its use in interpreting the mass spectrum. The ion *m/e* 41 requires that three carbon atoms are directly attached to each other and that they carry five hydrogen atoms. One simple interpretation of this is that the ion is of the form $[CH_3CH{=}CH_2]^+$ and that it constitutes one end of the original molecule. The complementary ion, the base peak *m/e* 59, has the molecular formula $[C_2H_5NO]^+$. If the nitrogen and oxygen were present as a nitroso group, the ion corresponding to this group, *m/e* 30, or to its loss at *m/e* 71, should be very abundant in the spectrum. This is not so, and therefore the two heteroatoms are either attached to the same or to adjacent carbon atoms. A detailed analysis of all the possibilities is avoided if one recalls that the base peak is a re-arrangement ion, and that there is one double-bond equivalent in the parent molecular ion, as indeed there is in the re-arrangement product. Previous compounds, namely butanoic acid and 2-hexanone, had prominent re-arrangement ions formed by fission β to a keto group and a concomitant hydrogen migration. By analogy, one may postulate that the present compound contains a keto group attached on one side to a methylene group. Thus, the unknown is a primary acid amide of the form $RCH_2CH_2CONH_2$ which, upon ionization and fragmentation with re-arrangement,[22] yields $[CH_3CONH_2]^+$(*m/e* 59).

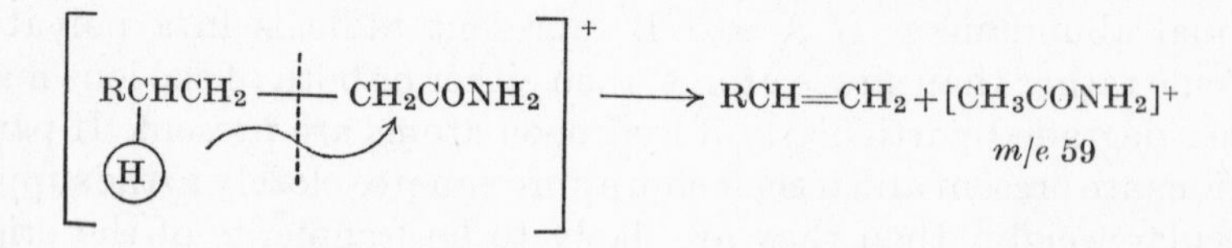

This is confirmed by the alternative fission $RCH_2CONH_2 \rightarrow [CONH_2]^+$ (*m/e* 44) + residue. Now the complete structure contains five carbon atoms, so that R′ in the formula $R'CH_2CH_2CONH_2$ is either a n-propyl or an isopropyl group. Clearly, it is difficult for an isopropyl substituent to lose an ethyl group without re-arrangement, whereas an n-propyl might reasonably do so. Moreover, as has already been mentioned, a structure containing an isopropyl group, such as isovaleramide, might be expected to yield an abundant ion *m/e* 43, where experience shows that a normal alkyl chain would yield the ions *m/e* 43 and 41 in comparable abundance. The present spectrum shows an abundant ion *m/e* 29, which is probably an ethyl group, and the two ions (43 and 41) to be of comparable abundance. The balance of the evidence, therefore, favours a n-alkyl chain and the molecule is probably n-valeramide.

It is useful to re-examine the spectrum and to assign probable formulas to each major fragment ion. Much of this has already been done in the discussion, and it merely remains to summarize the conclusions. The abundant fragment ions with their probable formulae are given in Table 11.

TABLE 11

m/e	Probable formulas	*m/e*	Probable formulas
		For the less abundant ions	
72	$CH_2CH_2CONH_2^+$	101	$C_4H_9CONH_2^+$
59	$CH_3CONH_2^+$	100	$C_4H_9CONH^+$
57	$CH_3CH_2CH_2CH_2^+$	86	$C_3H_6CONH_2^+$
44	$CONH_2^+$	85	$C_3H_6CONH^+$
43	$C_3H_7^+$	73	$C_2H_5CONH_2^+$
41	$C_3H_5^+$	60	$^{13}CH_3CONH_2^+$
29	$C_2H_5^+$		with $^{12}CH_3CO^{15}NH_2^+$
		55	$C_4H_7^+$
		42	$C_3H_6^+$
		28	$C_2H_4^+$ with CO^+
		27	$C_2H_3^+$
		15	CH_3^+

The ready assignment of all the fragment ions suggests that the proposed structure is correct.

The lack of ions arising from the presence of the isotopes ^{13}C, ^{15}N and ^{18}O and by means of which the ion *m/e* 101 could have been identified as a parent molecular ion, is a severe handicap in the analysis of this spectrum.* With the added, valuable, information provided by precise mass measurement the analysis is still of great difficulty. Since the alternative structure which was possible on the mass measurements was a fragment ion, one cannot be sure that some of the ions observed at lower mass numbers may not be other fragments of an unknown molecule of greater complexity which does not give rise to a parent molecular ion.

* In this spectrum, however, as in numbers 6, 7 and 9, reference to the *z*-number table, p. 127, would have helped considerably. [Ed.]

Accordingly, one can proceed only tentatively, as has been done. The correlation of the observed ions with the assumed structure provided strong confirmatory evidence of the correctness of the assumptions made.

TABLE 12

m/e	% Ab. (I)	$\frac{100 \times I}{\Sigma}$	m/e	% Ab. (I)	$\frac{100 \times I}{\Sigma}$
15	0·96	0·38	57	11·09	4·41
27	8·01	3·18	58	0·71	0·28
28	5·03	2·00	59	100·00	39·73
29	14·42	5·73	60	3·00	1·19
30	1·00	0·40	72	18·97	7·54
31	0·62	0·25	73	3·00	1·19
41	13·14	5·22	74	0·10	0·04
42	4·98	1·98	85	3·00	1·19
43	11·03	4·38	86	3·00	1·19
44	38·01	15·10	87	0·21	0·08
45	1·99	0·79	100	0·61	0·24
55	5·99	2·38	101	0·81	0·32
56	1·99	0·79			

$\Sigma = 251{\cdot}67$

References

1. Sharkey, A. G., Shultz, J. L. and Friedel, R. A. (1956). *Analyt. Chem.* **28**, 934.
2. Gilpin, J. A. and McLafferty, F. W. (1957). *Analyt. Chem.* **29**, 990.
3. Beynon, J. H. (1960). "Mass Spectrometry and its Applications to Organic Chemistry". Elsevier, Amsterdam. (a) p. 330, (b) p. 362, (c) p. 354.
4. Bloom, E. G., Mohler, F. L., Lengel, J. H. and Wise, C. E. (1948). *J. Res. nat. Bur. Stand.* **41**, 129.
5. Rylander, P. N., Meyerson, S. and Grubb, H. M. (1957). *J. Amer. chem. Soc.* **79**, 842.
6. Azcel, T. and Lumpkin, H. E. (1961). *Analyt. Chem.* **33**, 386.
7. Friedel, R. A. Shultz, J. L. and Sharkey, A. G. (1956). *Analyt. Chem.* **28**, 926.
8. Happ, G. P. and Stewart, D. W. (1952). *J. Amer. chem. Soc.* **74**, 4404.
9. Quayle, A. H. (1959). Colloq. Spec. Int. VIII, 259.
10. Sharkey, A. G., Shultz, J. L. and Friedel, R. A. (1959). *Analyt. Chem.* **31**, 87.
11. McLafferty, F. W. (1957). *Analyt. Chem.* **29**, 1782.
12. Peard, W. J. and McLafferty, F. W. (1959). A.S.T.M. E-14 Conference.
13. A.P.I. Catalogue of Mass Spectral Data, Research Project 44.
14. Field, F. H. and Franklin, J. L. (1957). "Electron Impact Phenomena and the Properties of Gaseous Ions". Academic Press, New York.
15. McLafferty, F. W. (1960). A.S.T.M. Committee E-14 on Mass Spectrometry Atlantic City.
16. McLafferty, F. W. (1957). *Appl. Spectrosc.* **11**, 148.
17. McLafferty, F. W. (1959). *Analyt. Chem.* **31**, 82.
18. Hipple, J. A., Fox, R. E. and Condon, E. O. (1946). *Phys. Rev.* **69**, 347.
19. Beynon, J. H. (1959). "Advances in Mass Spectrometry". p. 328. Pergamon Press, London.
20. Reed, R. I. and Wilson, J. M. (1962). *Chem. & Ind.* 1428.
21. Levy, E. J. and Stahl, W. H. (1957). A.S.T.M. E-14 Committee on Mass Spectrometry, New York.
22. Gilpin, J. A. (1959). *Analyt. Chem.* **31**, 935.

Bibliography

Biemann, K. (1962). "Mass Spectrometry. Organic Chemical Applications". McGraw-Hill, London.

Budzikiewicz, H., Djerassi C. and Williams D. H. (1964). "Structure Elucidation of Natural Products by Mass Spectrometry", Vol. 1. Alkaloids. Holden-Day, London.

Budzikiewicz, H., Djerassi, C. and Williams, D. H. (1964). "Interpretation of Mass Spectra of Organic Compounds". Holden-Day, San-Franciso.

McLafferty, F. W. (1963). "Mass Spectrometry of Organic Ions". Academic Press, New York.

"Mass Spectral Correlations". Advances in Chemistry Series No. 40 (1963). American Chemical Society.

Key to Unknown Spectra

Nuclear Magnetic Resonance

Unknown 1.

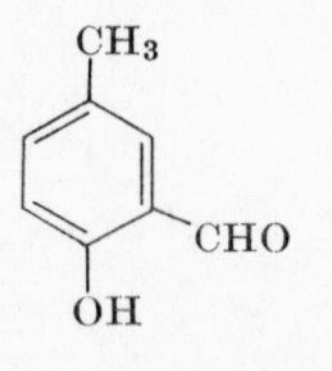

Unknown 2.

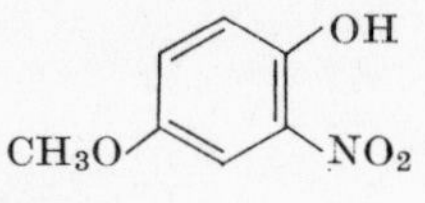

Unknown 3. Hexenolactone.

Unknown 4.

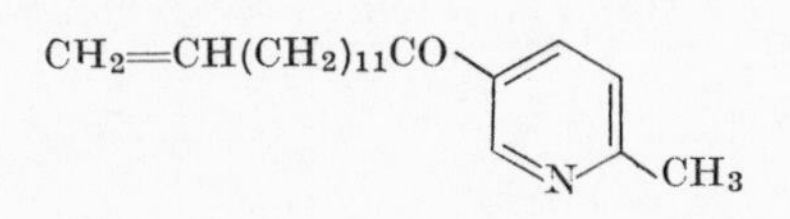

Unknown 5.

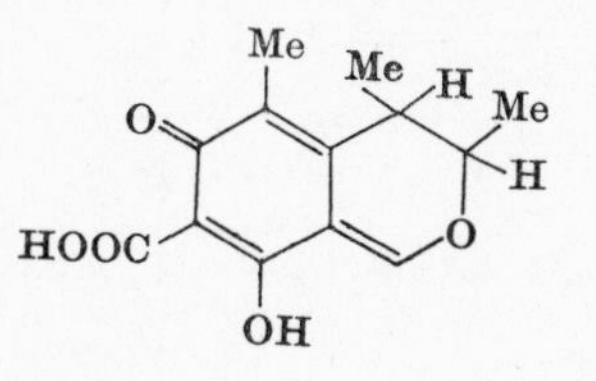

Infrared

Unknown A. Long-chain alkene with terminal double bond.

Unknown B. Di-n-butyl ether.

Unknown C. 1,2,4-Trimethylbenzene.

Unknown D. Acetanilide.

Unknown E. *p*-Chloroaniline.

Unknown F. α-Alanine.